Multivariate Analysis in Management, Engineering and the Sciences

Multivariate Analysis in Management, Engineering and the Sciences

Editor

Beata Akselsen

scitus academics

Multivariate Analysis in Management, Engineering and the Sciences
Edited by **Beata Akselsen**

ISBN: 978-1-68117-462-4
Library of Congress Control Number: 2016942824

www.scitusacademics.com
Box No. 4766, 616 Corporate Way,
Suite 2, Valley Cottage,
NY 10989

Notice

Printed in United States of America on Acid Free Paper

Preface

Many statistical techniques focus on just one or two variables; Multivariate analysis (MVA) techniques allow more than two variables to be analysed at once. Recently statistical knowledge has become an important requirement and occupies a prominent position in the exercise of various professions. In the real world, the processes have a large volume of data and are naturally multivariate and as such, require a proper treatment. For these conditions it is difficult or practically impossible to use methods of univariate statistics. Researchers use multivariate procedures in studies that involve more than one dependent variable (also known as the outcome or phenomenon of interest), more than one independent variable (also known as a predictor) or both. This type of analysis is desirable because researchers often hypothesize that a given outcome of interest is effected or influenced by more than one thing. Uses for multivariate analysis include: design for capability (also known as capability-based design); inverse design, where any variable can be treated as an independent variable; analysis of alternatives (AoA), the selection of concepts to fulfil a customer need; analysis of concepts with respect to changing scenarios; identification of critical design-drivers and correlations across hierarchical levels.

*Multivariate Analysis in Management, Engineering and the Sciences*presents significant topics on fundamental theoretical aspects of the field as well as on other aspects concerned with significant applications of new theoretical methods.Through real-life applications of statistical methodology, this book elucidates the implications of behavioural science studies for statistical analysis.In addition to helping to stimulate research in multivariate analysis, the book aims to bring about interactions among mathematical statisticians, probabilists, and scientists in other disciplines broadly interested in the area.

Table of Contents

1

CHAPTER

Application of Multivariate Statistical Analysis Concepts for Assessment of Hydrogeochemistry of Groundwater—A Study in Suri I snd II Blocks of Birbhum District, West Bengal, India

Shreya Das, , S. K. Nag*

Department of Geological Sciences, Jadavpur University

ABSTRACT

Multivariate statistical techniques, cluster and principal component analysis were applied to the data on groundwater quality of Suri I and II Blocks of Birbhum District, West Bengal, India, to extract principal factors corresponding to the different sources of variation in the hydrochemistry as well as the main controls on the hydrochemistry. For this, bore well water samples have been collected in two phases, during Post-monsoon (November 2012) and Pre-monsoon (April 2013) from 26 sampling locations spread homogeneously over the two blocks. Excess fluoride in groundwater has been reported at two locations both in post- and in pre-monsoon sessions, with a rise observed in pre-monsoon. Localized presence of excess iron has also been observed during both sessions. The water is found to be mildly alkaline in post-monsoon but slightly acidic at some locations during pre-monsoon. Correlation and cluster analysis studies demonstrate that fluoride shares a moderately positive correlation with pH in post-monsoon and a very strong one with carbonate in pre-monsoon indicating dominance of rock water interaction and ion exchange activity in the study area. Certain locations in the study area have been reported with less than 0.6 mg/l fluoride in groundwater, leading to possibility of occurrence of severe dental caries especially in children. Low values of sulfate and phosphate in water indicate a meager chance of contamination of groundwater due to anthropogenic factors.

Keywords: Groundwater quality Fluoride Correlation analysis Hierarchical cluster analysis Principal component and factor analysis Suri I and II

INTRODUCTION

Assessment of the hydrochemical characteristics of water and aquifer hydraulic properties is important for groundwater planning and management in the study area. Generally, the motion of groundwater along its flow paths below the ground surface increases the concentration of the chemical species (Domenico and Schwartz 1990; Freeze and Cherry 1979; Kortasi 2007). Groundwater is the purest form of water sourced from natural resources and is usually clear, colorless and remains relatively at constant temperature. Therefore, it is normally superior to surface water in terms of sanitary consideration. But groundwater has higher salt contents than surface water because slowly moving water remains in contact with sub-strata for longer periods. Hence, the groundwater chemistry could reveal important information on the geological history of the aquifers as well as the suitability of groundwater for domestic and agricultural purposes. Quality of groundwater is equally important to its quantity owing to the suitability of water for various purposes (Schiavo et al. 2006; Subramani et al.2005). Hydrochemical evaluation of groundwater systems is usually based on the availability of a large amount of information concerning groundwater chemistry (Aghazadeh and Mogaddam 2010; Hossein 2004). Groundwater chemistry, in turn, depends on a number of factors, such as general geology, degree of chemical weathering of the various rock types, quality of recharge water and inputs from sources other than water rock interaction. Such factors and their interactions result in a complex groundwater quality (Domenico and Schwartz 1990; Guler and Thyne2004; Sunne et al. 2005). Various studies on the hydrogeological and hydrochemical properties of the aquifers underlying the country are currently being pursued. Recent studies have been focused on the quality of groundwater from specific hydrogeologic terrains in the country. Researchers (e.g. Kortatsi et al. 2008; Yidana et al. 2008a, b,c, d; Yidana 2008; Banoeng-Yakubo et al. 2009; Nag and Ray 2015; Nag 2014; Nag and Ghosh 2013) have applied different methodologies to understand the sources of variation in the quality of surface and groundwater basins in India.

The competition for water resources has gained importance in recent years, not only in India but also in many places of the world. The development of human societies and industry result in bioenvironmental problems;

pollution puts the water, air and soil resources at risk (Milovanovic 2007). Groundwater has become the major source of water supply for domestic, industrial and agricultural sectors of many countries and groundwater chemistry depends on the quality of recharged water, atmospheric precipitation, inland surface water and sub-surface geochemical processes. In recent years, many cities of developing countries are experiencing rapid demographic growth due to rural exodus. Urbanization and the unregulated growth of the population have altered the local topography and drainage system directly which affect both quality and quantity of the groundwater (Vasanthavigar et al. 2010). Temporal changes in the origin and constitution of the recharged water, hydrological and human factors frequently cause periodic changes in groundwater chemistry and quality (Milovanovic2007; Sreedevi 2004).

The present study has been carried out in two blocks—Suri I and II, of Birbhum district, West Bengal, India. Groundwater is an important water resource for drinking, agriculture and industrial use in the present study area. In this study, physical and hydrochemical data from the groundwater system has been used to determine the main factors and mechanisms controlling the chemistry of groundwater in the area. Geochemical studies of groundwater provide a better understanding of possible changes in quality as development progress. Besides anthropogenic factors, natural factors also play a role in contaminating groundwater, rendering it unfit for drinking. Fluoride contamination in groundwater is one such factor which leads to irreparable physiological damage to the human body in form of dental or skeletal fluorosis. As per World Health Organization guidelines, the permissible limit of fluoride in drinking water is 1.5 mg/l (WHO 2011). Intense agricultural and urban development has caused a high rise in demand of groundwater resources in arid and semi-arid regions of West Bengal and thereby the resource has been put at a greater risk to contamination (Moghaddam and Najib 2006; Jalali 2006; Khazaei et al. 2006). In the present study area, people depend solely on groundwater supply for all domestic chores as well as for consumption. Excess fluoride in groundwater has been reported previously in Junidpur, Rampurhat and Nalhati blocks of Birbhum district (Gupta et al. 2006; Gupta and Banerjee 2011). The present study reports the presence of excess fluoride in groundwater and intends to identify the hydrogeochemical and anthropogenic contamination factors responsible for deterioration of water quality using statistical concepts.

Inter-relationships amongst the analyzed water quality parameters, with special emphasis on relation of fluoride with the rest of the parameters have been studied with the use of multivariate statistical analysis tools in this study. In the analysis of hydrochemical data, the application of multivariate statistical analyses has been profusely taken up by various researchers. These analyses assist us in systematically studying and interpreting hydrogeochemical patterns and correlationships between the various parameters determining water quality. Some examples of studies were multivariate analyses methods have been effectively employed in comprehending hydrochemical processes are studies by Güler et al. (2002), Güler and Thyne (2004), Helsrup et al. (2007), Yidana et al. (2008a), Yidana and Yidana (2009), Cloutier et al. (2008), Machiwal and Jha (2015), Mrazovac and Miloradov (2011), Gummadi et al. 2014, Bajpayee et al. 2012, Jothivenkatachalam et al. 2010 and others. Multivariate statistical methods do not directly present hydrogeochemical processes being carried out but gives an idea about the factors that control these processes.

STUDY AREA

The present study has been carried out in Suri (comprising of two blocks—Suri I and Suri II), the district headquarter of Birbhum district, West Bengal, India. The two blocks are located between latitudes 23.76°N–23.99°N and longitudes 87.42°E–87.64°E, and covers an area of 220 km^2 (approx.). The area usually experiences an arid climate. Summer temperatures soar to a maximum of 40 °C or above whereas in winter temperatures dip to around 10 °C. Majority of the rainfall is limited to the monsoon season from June to October and hovers around an average of 1100 mm. The area is characterized by rural setting and major occupation of the people is agriculture. The study area is largely comprised of alternating layers of sand and clay, which are soft sediments and part of the Ganga-Kosi formation and are of the tertiary age. The clays interbedded with sand partings are poorly sorted with occasional layers of grit. Granite gneiss which date back to the Archean age, are hard and foliated type rocks belonging to the Chotanagpur Gneissic complex and, constitute the north western part of the study area (Mukherjee et al. 1969). These poly-metamorphosed rocks contain streaks of pegmatite, alkali feldspars and mica. Hard clays dominate small areas in the extreme upper and lower parts and a stretch in the central part-all lying in the eastern region of the study area. Presence of laterites and lateritic soil, which render a reddish color to the top soil in many regions of the study area, is observed to have a scattered distribution spanning both Suri I and II blocks. Figure 1presents the study area map marked with the 26 sampling location points from where water samples have been collected along with the geological map of the study area.

MATERIALS AND METHODS

Groundwater sample collection from the study area has been done in two phases: post-monsoon and pre-monsoon sessions. Water samples were collected from bore to wells at 26 sampling locations spread homogeneously over the two blocks, Suri I and II. Depth of the water table ranged widely in the study area. During post-monsoon the Depth to Water Level (DTWL) values ranged from 1.69 to 13.83 m and during pre-monsoon it ranged from 3.03 to 21.24 m. The samples were collected in 500 ml sterilized polythene bottles. For all samples collected, parameters like pH, electrical conductivity (EC) and total dissolved solids (TDS) were measured in the field. This is necessary as they are likely to change during transport. The water samples were tested for a total of sixteen parameters. The quantitative analysis for major cations and anions in water including fluoride was carried out in the laboratory employing standard methods (APHA 1995). Various statistical analysis methods have been used in the past to determine water quality and hydrogeochemistry (Nelson and Ward 1981; Briz-Kishore and Murali 1992; Ramesh Kumar and Riyazuddin 2008; Routroy et al. 2013). Based on the quantitative chemical analysis data generated, such statistical analyses to the likes of correlation analysis, hierarchical cluster analysis, principal component and factor analysis were performed using software program StatistiXL.

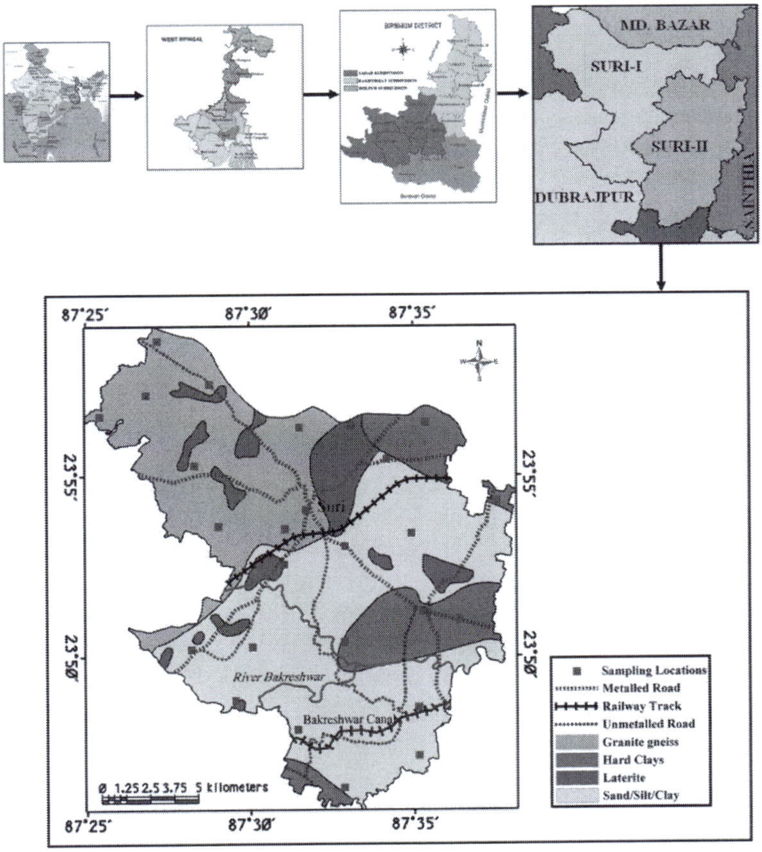

Figure 1. Map of the study area presenting the sampling location points and geology of the area

RESULTS AND DISCUSSIONS

Major ion chemistry and spatial distribution

The pH values of the groundwater varies from 7.00 to 8.40 (in post-monsoon) with an average of 7.50, and 6.4–8.4 (in pre-monsoon) with an average of 7.1, which indicates that water is almost neutral in nature. The average concentration of Total Dissolved Solids (TDS) ranged from 244.3 (post-monsoon) to 249.0 (pre-monsoon) mg/l in the study area. Normally TDS in water may originate from natural sources and sewage discharges. The electrical conductivity (EC) in the study area varies from 90.00 to 300.00 (in post-monsoon) and 55.00 to 590.00 (in pre-monsoon) with an average of 212.7 (post-monsoon) and 556.2 (pre-monsoon) µS/cm at 25 °C. The Total hardness in water is derived from the solution of carbon dioxide released by bacterial action in the soil, in percolating rain water. Low pH conditions develop and lead to the dissolution of insoluble carbonates in the soil and in limestone formations to convert them into soluble bicarbonates. Impurities in limestone, such as sulfates, chlorides and silicates, become exposed to the solvent action of water as the carbonates are dissolved so that they also pass into solution. The total hardness (TH) of water is a measure of mainly calcium carbonate and magnesium carbonate dissolved in groundwater. The general acceptance level of hardness is 300 mg/l, although WHO has set an allowable limit of 600 mg/l. The total hardness in the study area ranges between 55 and 365 mg/l in post-monsoon while in pre-monsoon it ranges between 48 and 384 mg/l.

Calcium concentration ranged from 12.6 to 109.2 mg/l in post-monsoon and 6.7–95.8 mg/l in pre-monsoon periods. The magnesium concentration ranges between 3.2–42.5 mg/l in post-monsoon and 7.6–35.3 mg/l. in pre-monsoon. Sodium concentration in water varies from 7.00 to 58.3 mg/l with an average of 26.04 mg/l in post-monsoon and 8.0–69.0 mg/l with an average of 30.5 mg/l in pre-monsoon period. Iron is an essential element in the human body and is required physiologically on various aspects (Moore 1973). Although iron has little concern as a health hazard, it is still considered as a nuisance in excessive quantities (Dart 1974). It causes staining of clothes and utensils. It is also not suitable for processing of food, beverages, dyeing, bleaching, etc. The concentration limits of iron in drinking water ranges between 0.3 mg/l (maximum acceptable) and 1.0 mg/l (maximum allowable). Iron concentrations of Suri I and II blocks range between 0.0 and 1.8 mg/l in post-monsoon and 0.0 to 6.1 mg/l in pre-monsoon. At Bhagabanbati Primary School (Location no. 13 located in the eastern part of the study area) the iron concentration is above the desirable limit (0.30 mg/l) during both post and pre-monsoon. High iron concentration affects the taste of water, has adverse effects on domestic uses and promotes growth of iron bacteria. Measures should be taken before consumption by installation of iron removing plants.

Bicarbonate ion varies from 48.80 to 1073.60 and 48.8 to 292.8 mg/l in post and pre-monsoon respectively. Chloride concentration in groundwater samples in the study area ranged from 15.0 to 124.96 and 15.0 to 274.9 mg/l in post and pre-monsoon respectively. WHO has set standards of 200–500 mg/l for chloride in drinking water. Too much of chloride leads to bad taste in water and also chloride ion combines with the Na (that is being derived from the weathering of granitic terrains) and forms NaCl, whose excess presence in water makes it saline and unfit for drinking and irrigation purposes. Here too, as exhibited by mean values in Table 1, the chloride value decreases during post-monsoon.

Table 1. Quantitative chemical analysis results of post and pre-monsoon period

Parameter	Post-monsoon						Pre-monsoon					
	Minimum	Maximum	Mean	Std. Deviation	Skewness	Kurtosis	Minimum	Maximum	Mean	Std. Deviation	Skewness	Kurtosis
pH	7.0	8.4	7.5	0.34	0.37	0.09	6.4	8.4	7.1	0.43	0.87	1.52
TDS	74.0	622.0	244.3	109.2	1.41	4.81	55.0	590.0	249.0	125.9	0.76	1.04
EC	90.0	300.0	212.7	48.65	−0.54	1.01	150.0	1200.0	556.2	256.6	1.05	1.80
TA	50.0	880.0	530.0	204.24	−0.54	0.15	40.0	320.0	158.8	66.7	0.42	0.10
TH	55.0	365.0	206.9	70.5	−0.07	0.17	48.0	384.0	148.3	68.7	1.66	4.52
Ca^{2+}	12.6	109.2	46.8	21.1	0.97	1.81	6.7	95.8	29.3	18.4	2.11	5.87
Mg^{2+}	3.2	42.5	21.9	8.0	0.06	1.12	7.6	35.3	18.3	7.9	0.36	−0.76
Na^{+}	7.0	58.3	26.04	15.4	0.80	−0.38	8.0	69.0	30.5	17.8	0.81	−0.20
K^{+}	0.7	10.0	2.4	2.1	2.22	6.49	0.2	2.5	0.6	0.54	2.00	4.69
Fe^{2+}	0.0	1.8	0.12	0.35	4.95	24.94	0.0	6.1	0.47	1.21	4.39	20.33
CO_3^{2-}	0.0	204.0	56.8	75.5	1.07	−0.48	0.0	36.0	6.0	9.75	1.70	2.52
HCO_3^{-}	48.8	1073.6	531.2	222.6	−0.36	1.22	48.8	390.4	181.6	76.2	0.75	1.10
Cl^{-}	15.0	129.9	39.4	31.15	1.99	3.52	15.0	274.9	49.0	58.65	2.93	9.16
SO_4^{2-}	0.43	48.8	8.4	13.43	−0.36	1.22	0.1	56.7	8.7	15.03	2.43	4.94
PO_4^{3-}	0.0	0.04	0.01	0.01	1.23	1.59	0.0	4.8	0.3	0.93	4.80	23.80
F^{-}	0.21	2.38	0.6	0.50	2.36	6.34	0.24	2.84	0.73	0.59	2.34	6.44

The sulfate ion causes no particular harmful effects on soils or plants; however, it contributes in increasing the salinity in the soil solution. Sulfate ion varied from 0.43 to 48.76 mg/l during post-monsoon and 0.1 to 56.7 mg/l in pre-monsoon periods.

The range of presence of each of the sixteen parameters analyzed, along with their average and standard deviation values for both the sampling sessions have been presented in Table 2. Based on the analyses performed on the quantitative data, the following results have been reported.

Table 2. Correlation coefficient matrix for post-monsoon session

Parameters	PH	TDS	EC	TA	TH	Ca^{2+}	Mg^{2+}	Na^{+}	K^{+}	Fe^{2+}	CO_3^{2-}	HCO_3^{-}	Cl^{-}	SO_4^{2-}	PO_4^{3-}	F^{-}
pH	1.00															
TDS	–0.42	1.00														
EC	–0.36	0.53	1.00													
TA	–	0.55	0.59	1.00												
TH	–0.66	0.84	0.61	0.49	1.00											
Ca^{2+}	–0.71	0.85	0.45	0.43	0.92	1.00										
Mg^{2+}	–0.28	0.45	0.58	0.36	0.69	0.33	1.00									
Na^{+}	–	0.61	0.58	0.57	0.38	0.31	0.19	1.00								
K^{+}	–0.08	0.43	0.48	0.40	0.25	0.26	0.17	0.46	1.00							
Fe^{2+}	–0.15	0.11	0.13	–0.15	0.19	0.20	0.03	–	0.13	1.00						
CO_3^{2-}	0.19	0.04	0.43	0.47	–0.05	0.02	–0.15	0.47	0.38	–0.08	1.00					
HCO_3^{-}	–0.13	0.58	0.37	0.79	0.66	0.50	0.51	0.31	0.19	–0.11	–0.16	1.00				
Cl^{-}	–0.48	0.70	0.35	0.00	0.52	0.61	0.20	0.42	0.32	–0.02	–0.13	0.09	1.00			
SO_4^{2-}	–0.47	0.75	0.42	0.13	0.51	0.59	0.22	0.56	0.58	–0.08	0.01	0.14	0.85	1.00		
PO_4^{3-}	0.14	–0.16	–0.31	–0.24	–0.11	–0.14	0.11	–0.08	–0.09	0.07	–0.15	-0.17	–0.12	–0.02	1.00	
F^{-}	0.41	0.03	0.14	0.38	–0.14	–0.21	0.04	0.46	0.03	–0.10	0.22	0.27	–0.07	–0.16	–0.17	1.00

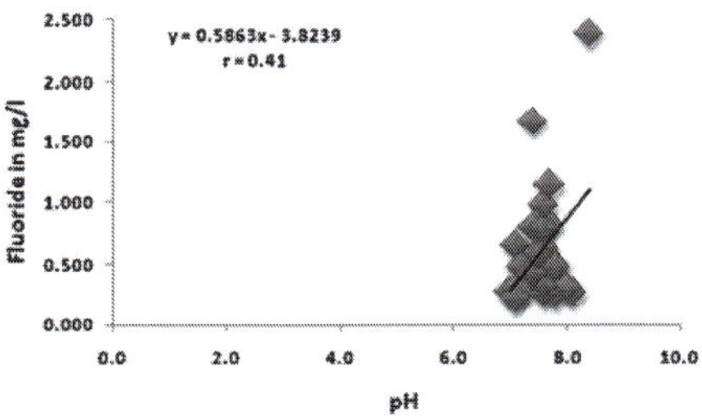

(a)

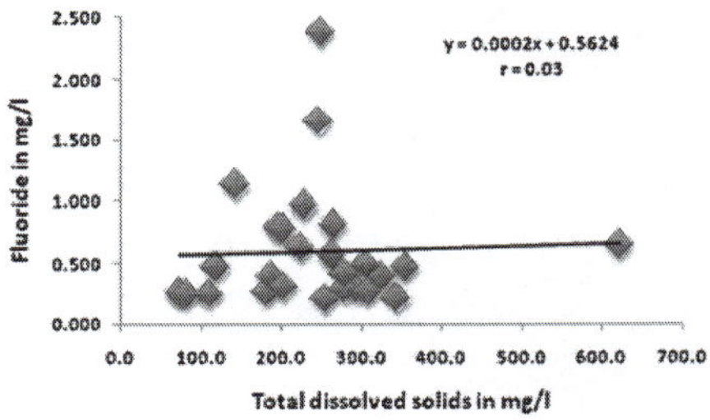

(b)

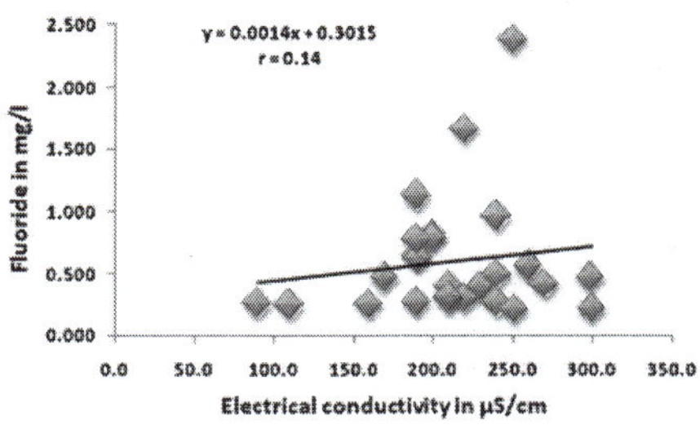

(c)

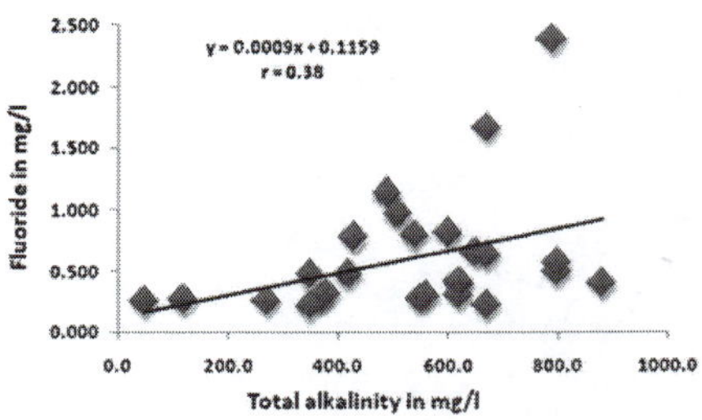

(d)

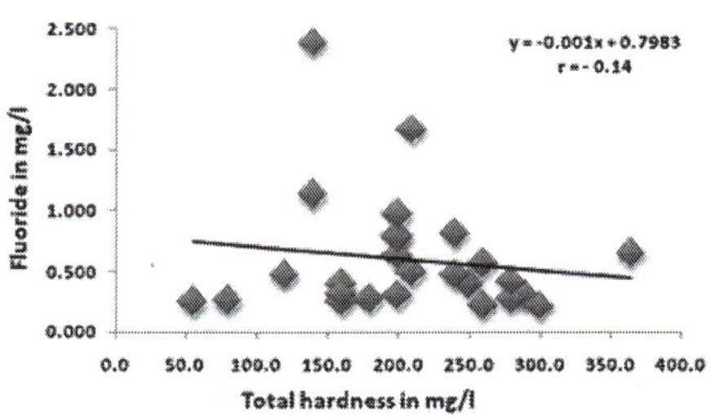

(e)

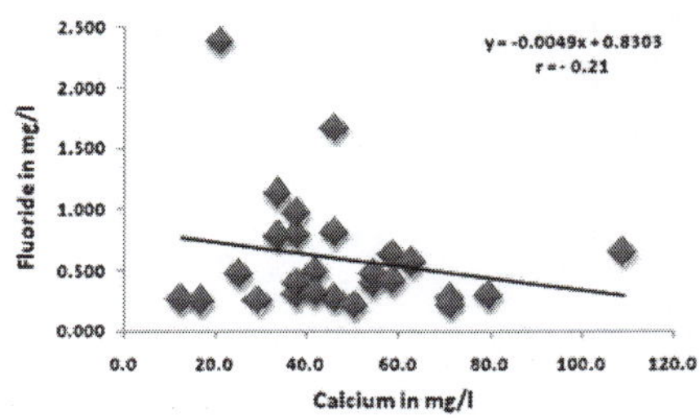

(f)

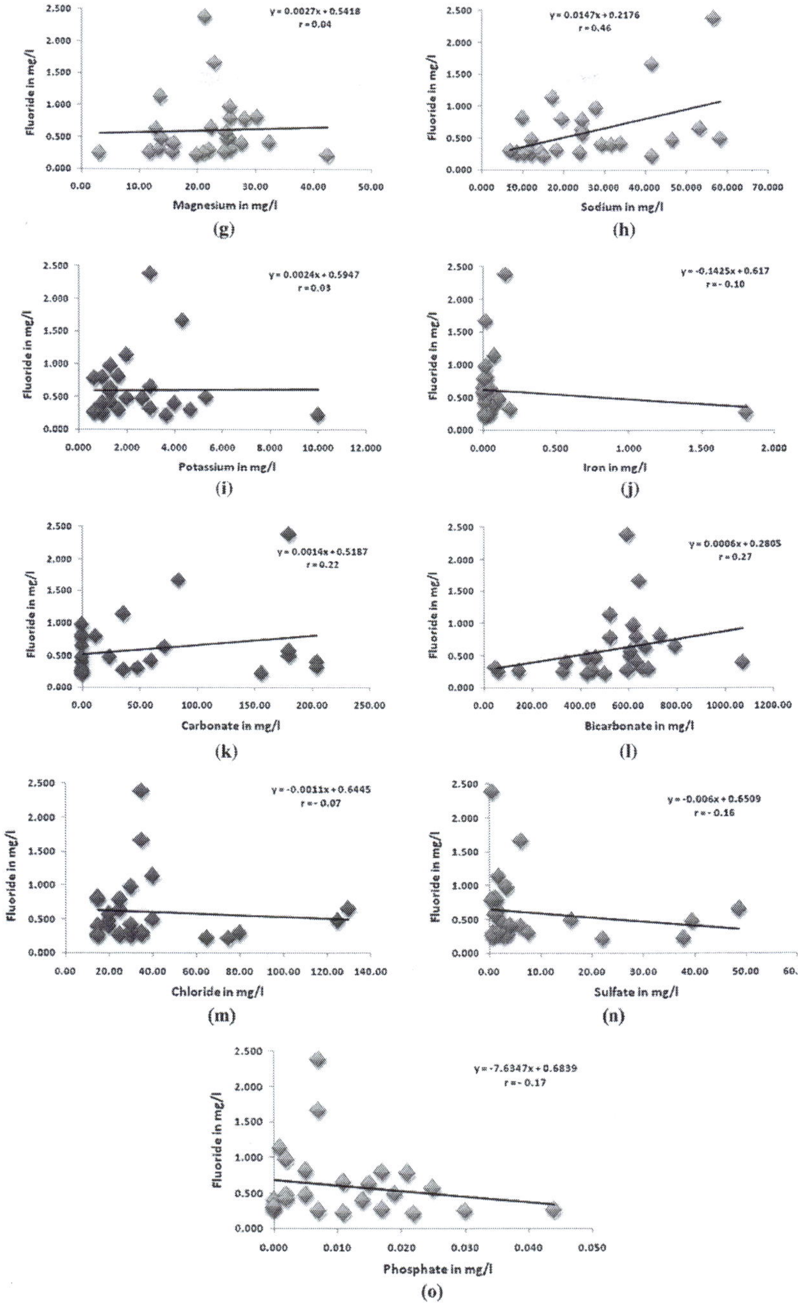

Figure 2. Correlation Analysis graphs presenting correlation trends of fluoride with the other analyzed water quality parameters in post-monsoon session

CORRELATION ANALYSIS

The correlation analysis study involving statistical calculations was devised by Pearson (1896). Based on the value of correlation coefficient '*r*', the correlation between two parameters plotted on a XY scatter diagram can be termed as positive or negative. Correlation analysis is a common and useful statistical tool for water quality studies indicating which ions control the water chemistry (Box et al. 1978; Chapman1996; Shrivastava and Patil 2002; Zeng et al. 2005). It is simply a measure to exhibit how well one variable predicts the other (Kurumbein and Graybill 1965). In the present study correlationship of fluoride has been worked out with rest of the analyzed water quality parameters and the following observations have been made from the trend analysis graphs for both post-monsoon (Fig. 2) and pre-monsoon (Fig. 3) sessions. Tables 3 and 4 present the correlation coefficient matrices prepared for the post-monsoon and pre-monsoon sessions, respectively.

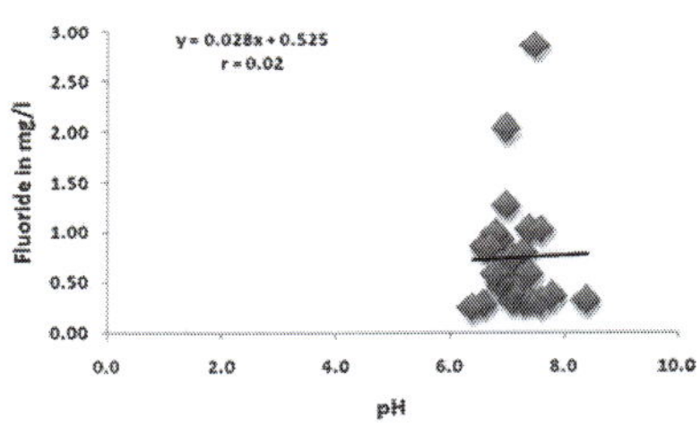

(a)

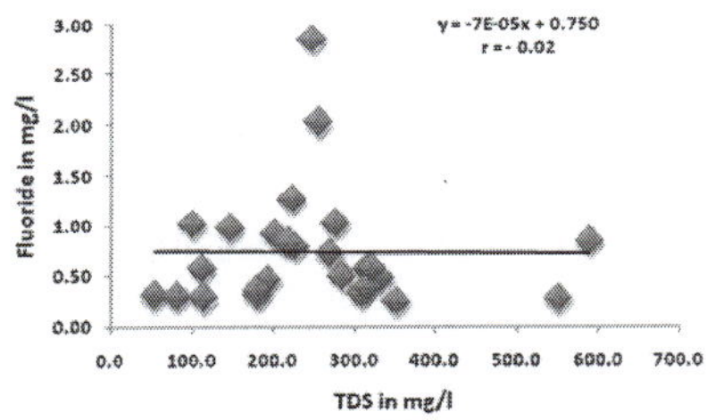

(b)

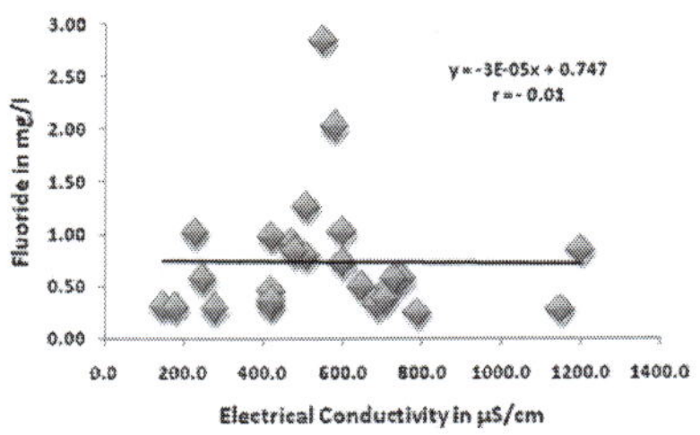

(c)

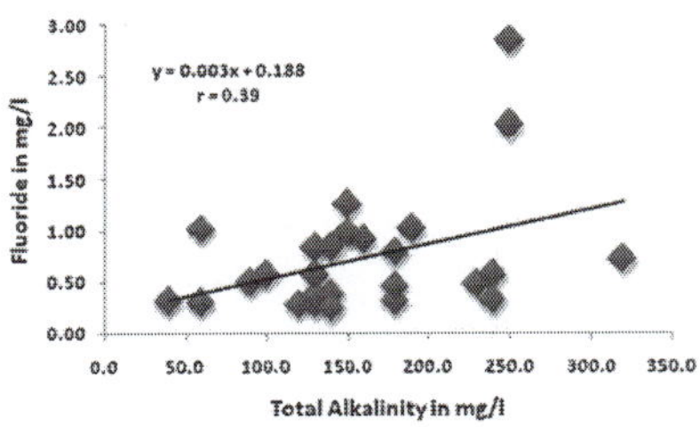

(d)

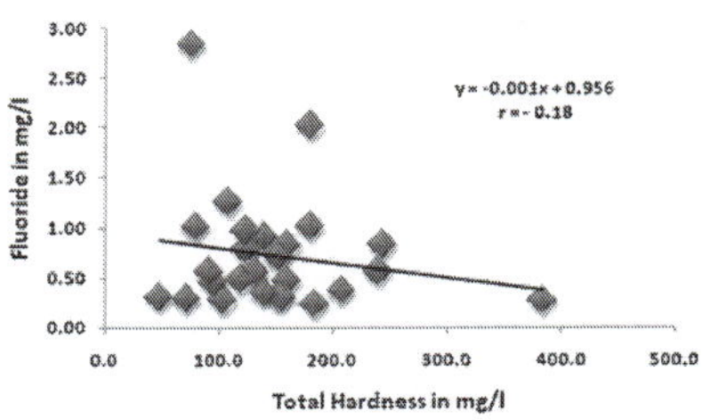

(e)

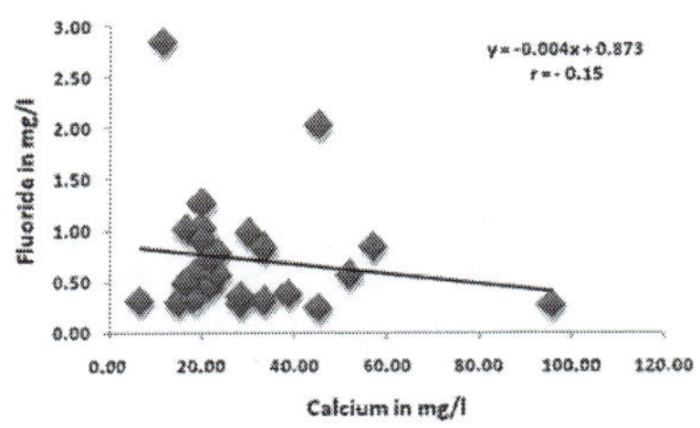

(f)

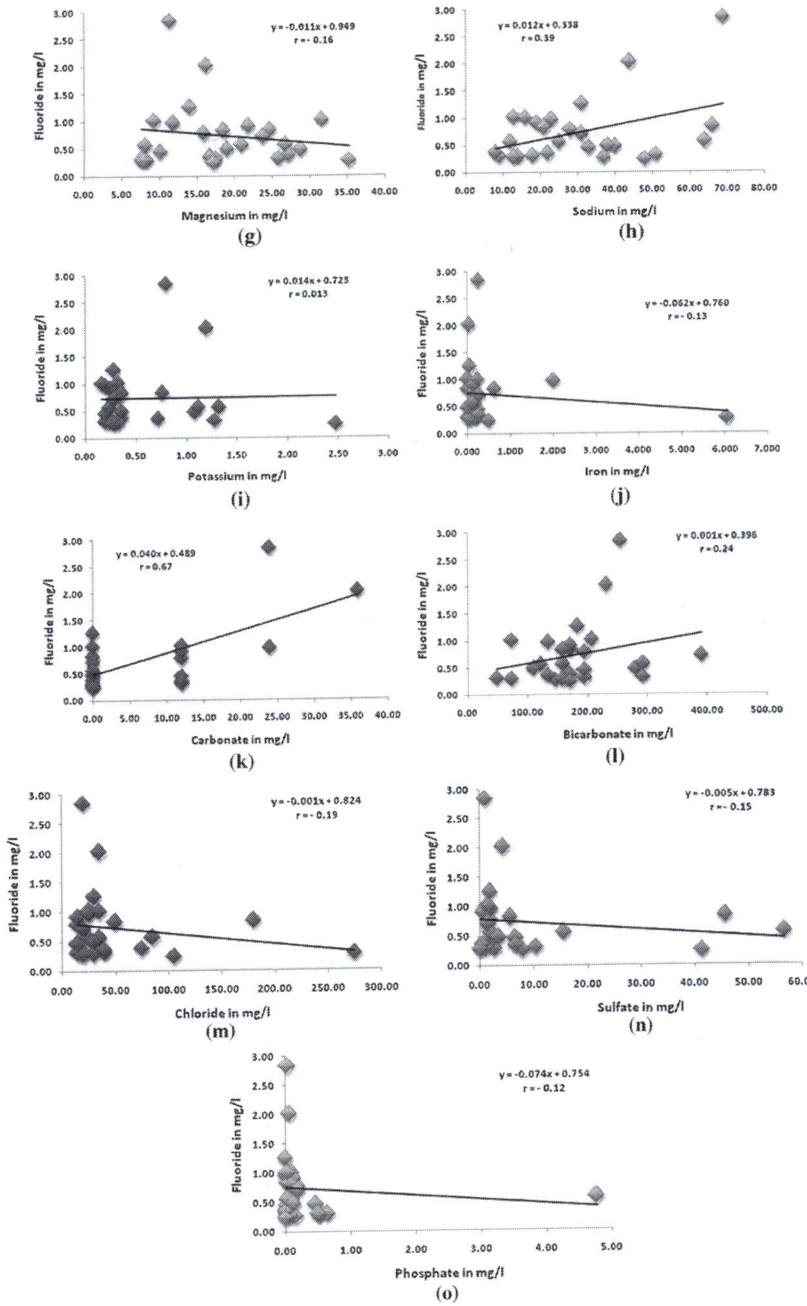

Figure 3. Correlation Analysis graphs presenting correlation trends of fluoride with the other analyzed water quality parameters in pre-monsoon session

Table 3. Correlation coefficient matrix for pre-monsoon session

Parameters	PH	TDS	EC	TA	TH	Ca^{2+}	Mg^{2+}	Na^{+}	K^{+}	Fe^{2+}	CO_3^{2-}	HCO_3^{-}	Cl^{-}	SO_4^{2-}	PO_4^{3-}	F^{-}
pH	1.00															
TDS	−0.56	1.00														
EC	−0.56	0.99	1.00													
TA	−0.10	0.32	0.29	1.00												
TH	−0.57	0.84	0.84	0.12	1.00											
Ca^{2+}	−0.60	0.69	0.70	−0.07	0.91	1.00										
Mg^{2+}	−0.35	0.79	0.78	0.36	0.82	0.51	1.00									
Na^{+}	−0.26	0.64	0.62	0.53	0.21	0.16	0.23	1.00								
K^{+}	−0.24	0.44	0.40	0.31	0.23	0.18	0.23	0.60	1.00							
Fe^{2+}	−0.33	0.42	0.42	−0.11	0.65	0.72	0.36	0.05	−0.12	1.00						
CO_3^{2-}	0.02	−0.14	−0.16	0.37	−0.11	−0.03	−0.18	0.15	−0.01	−0.01	1.00					
HCO_3^{-}	−0.12	0.38	0.36	0.97	0.16	−0.07	0.43	0.53	0.33	−0.11	0.14	1.00				
Cl^{-}	−0.49	0.76	0.79	−0.18	0.86	0.90	0.53	0.29	0.18	0.73	−0.26	−0.13	1.00			
SO_4^{2-}	−0.44	0.57	0.56	−0.06	0.47	0.47	0.32	0.40	0.64	−0.03	−0.26	0.01	0.49	1.00		
PO_4^{3-}	−0.08	0.08	0.06	−0.12	0.23	0.20	0.20	−0.12	0.16	−0.07	−0.16	−0.09	0.09	0.61	1.00	
F^{-}	0.02	−0.01	−0.02	0.39	−0.18	−0.15	−0.16	0.39	0.01	−0.13	0.67	0.24	−0.19	−0.15	−0.12	1.00

Table 4. Varimax rotated factor loadings for post-monsoon session

Variable	Factor 1	Factor 2	Factor 3	Factor 4	Factor 5	Factor 6
pH	0.478	0.065	−0.294	−0.209	0.237	−0.611
TDS	−0.745	0.151	0.555	0.101	−0.097	−0.085
EC	−0.272	0.632	0.513	0.201	−0.178	0.040
TA	−0.014	0.478	0.714	−0.133	−0.227	−0.334
TH	−0.517	0.054	0.779	0.190	−0.035	0.240
Ca^{2+}	−0.629	0.039	0.547	0.221	−0.223	0.258
Mg^{2+}	−0.101	0.054	0.799	0.053	0.284	0.102
Na^{+}	−0.524	0.521	0.194	0.081	0.029	−0.546
K^{+}	−0.401	0.663	0.130	−0.278	0.088	0.043
Fe^{2+}	−0.007	−0.030	0.004	0.948	0.055	0.046
CO_3^{2-}	0.116	0.899	-0.113	0.028	−0.125	−0.161
HCO_3^{-}	−0.096	−0.085	0.877	−0.169	−0.168	−0.263
Cl^{-}	−0.945	−0.029	0.049	−0.002	−0.058	0.047
SO_4^{2-}	−0.934	0.213	0.088	−0.119	0.090	0.103
PO_4^{3-}	0.053	−0.118	−0.026	0.050	0.932	0.048
F^{-}	0.074	0.076	0.113	−0.003	−0.132	−0.889
Eigen value	6.110	2.677	1.785	1.203	1.109	1.001
% Variance	38.19	16.73	11.16	7.52	6.93	6.26
Cumulative % variance	38.19	54.92	66.08	73.59	80.52	86.78

In the post-monsoon session, fluoride is observed to share a positive correlation with pH ($r = 0.41$), total alkalinity ($r = 0.38$), sodium ($r = 0.46$), carbonate and bicarbonate ($r > 0.2$). In case of magnesium, potassium and TDS the '*r*' value is positive but very close to zero indicating a not so strong correlation of these parameters with fluoride. Correlation of fluoride with calcium, total hardness, sulfate and phosphate is found to be negative. Bicarbonate shares positive correlations with alkalinity ($r = 0.79$), hardness ($r = 0.66$), calcium and magnesium ($r > 0.5$). The statistical significance of these correlations has been tested at p value = 0.05; and all have been found to be significant.

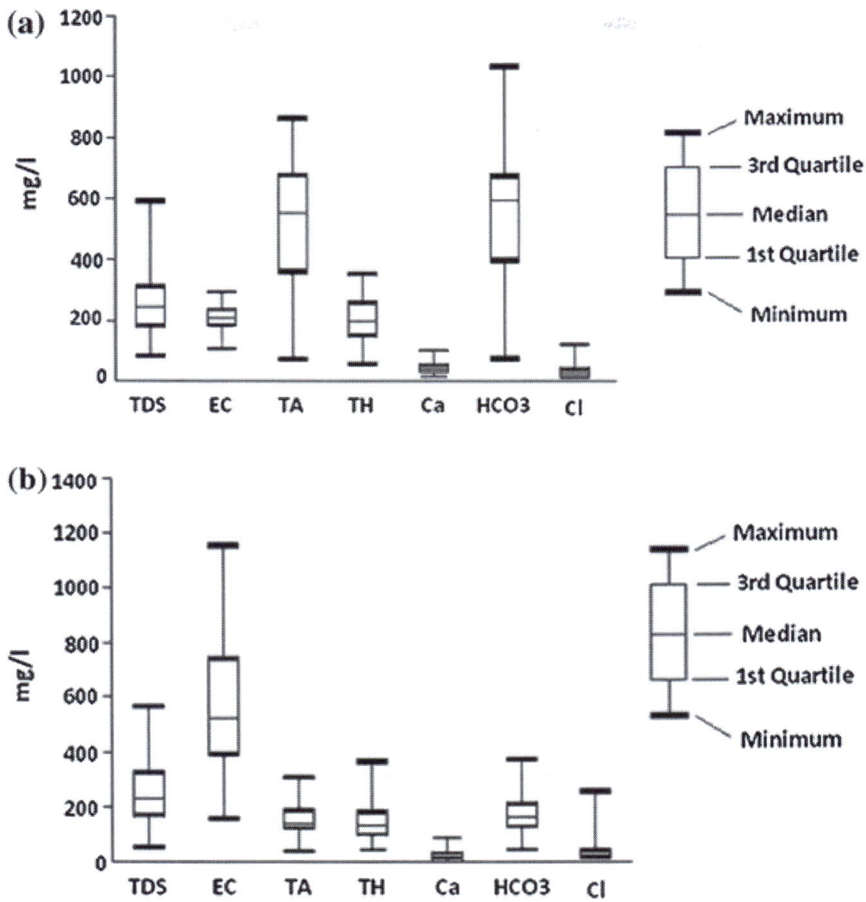

Figure 4. **a** *Box and Whisker Plot* for Post-monsoon session, **b** *box and Whisker Plot* for Pre-monsoon session

In the pre-monsoon session, fluoride shares fairly positive correlation with total alkalinity, sodium ($r = 0.39$), carbonate ($r = 0.67$) and bicarbonate. In case of correlation of fluoride with pH, although parameters are positively related, but the '*r*' value is very close to zero, a trend opposite than that observed during post-monsoon. With almost all the other analyzed parameters, fluoride shares a negative correlationship. The correlation of fluoride with potassium, iron and phosphate is found to be statistically insignificant during the pre-monsoon session. In the pre-monsoon session bicarbonate shares a strong but statistically insignificant correlationship ($r = 0.97$) with alkalinity whereas with hardness it is the opposite ($r = 0.16$). Both calcium and magnesium show strong correlations with total hardness ($r = 0.91$ and$r = 0.82$ respectively) indicating that Ca and Mg regulate the alkalinity character of groundwater. Iron shares a strong correlation with hardness ($r = 0.65$) as opposed to that during post-monsoon.

The box and whisker plots (Fig. 4a, b), representing a comparison of distribution of the analyzed water quality parameters based on maximum, minimum and quartile values, primarily indicate that groundwater in the study area is comparatively more alkaline with much lower conductivity during post-monsoon than that during pre-monsoon.

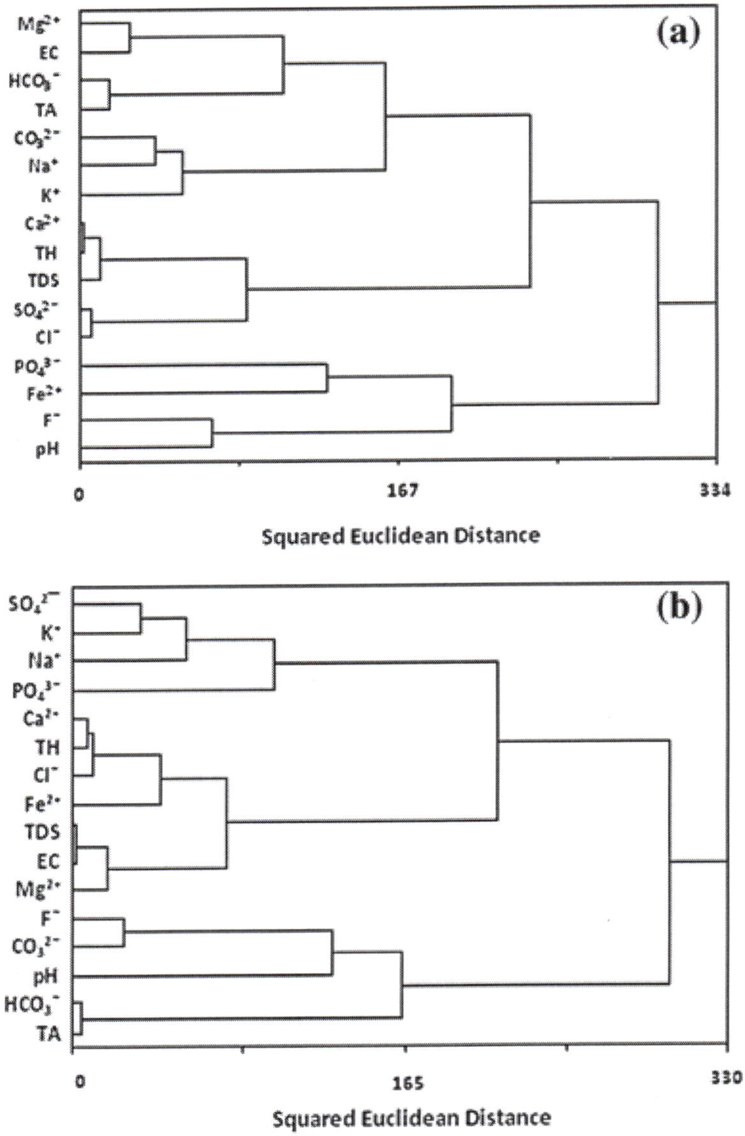

Figure 5. a Dendrogram generated through cluster analysis for post-monsoon session, **b** dendrogram generated through cluster analysis for pre-monsoon session

HIERARCHICAL CLUSTER ANALYSIS

Multivariate statistical analysis methods have been put to use in various water quality and hydrogeochemical studies. With the help of these analytical tools various forms of grouping studies can be performed in case of water samples collected from various sources. The multiple water quality parameters analyzed can be clustered based on their inter-dependency (Routroy et al. 2013) or sampling locations can be clustered based on chemical similarity in samples collected from these locations (Singh et al.2005; Shrestha and Kazama 2006). In the present study, the StatistiXL software has been used to run the Hierarchical cluster analysis function to group the multiple water quality parameters in clusters and generate dendrograms for the post-monsoon (Fig. 5a) and pre-monsoon (Fig. 5b) sessions. The dendrograms generated demonstrate high homogeneity within the clusters and high heterogeneity between the clusters (Sarbu and Pop 2005). The analyzed values of water quality parameters tested have been standardized (using the mean and standard deviation values) to lessen the disparity in range of the quantitative values of each parameter. The linkage distance amongst the parameters was measured in terms of squared euclidean distance (and the clustering method chosen was Ward's method.

In the post-monsoon session, the closest clusters generated were that of calcium-total hardness and sulfate-chloride. The alkalinity or salinity character of water is depicted by the sodium-carbonate and bicarbonate-total alkalinity clusters which further group with the magnesium—conductivity cluster. Fluoride forms cluster with pH and is primarily linked to the iron-phosphate cluster. On a larger scale three major clusters can be observed from top—the first one comprising seven parameters (Mg^{2+}, EC, HCO_3^-, TA, CO_3^{2-}, Na^+, K^+), second one comprising five parameters (Ca^{2+}, TH, TDS, SO_4^{2-}, Cl^-) and the third one of four parameters (PO_4^{3-}, Fe^{2+}, F^-, pH).

In the pre-monsoon session, the hardness factors like calcium, magnesium, total hardness are clubbed with TDS and conductivity. Fluoride is clustered with the alkaline parameters such as pH, carbonate, bicarbonate and total alkalinity. Sulfate and phosphate form another distinct cluster in the pre-monsoon session demonstrating the factor of anthropogenic contamination of groundwater. In form of larger clusters three major ones can be observed from top—comprising of four (SO_4^{2-}, K^+, Na^+, PO_4^{3-}), seven (Ca^{2+}, TH, Cl^-, Fe^{2+}, TDS, EC, Mg^{2+}) and five parameters (F^-, CO_3^{2-}, pH, HCO_3^-, TA).

The major variation observed in clustering patterns of both sessions is mainly related to fluoride. In post-monsoon session, fluoride contamination seems to be regulated by iron concentration and pH of groundwater whereas in pre-monsoon fluoride contamination in groundwater seems to be regulated primarily by the alkaline factors. Iron in the pre-monsoon session groups with hardness factors. Lateritic soil and clay soils dominate parts of the present study area as depicted in the geological map. Hence localized presence of iron in excess amounts in groundwater has been observed both during post and pre-monsoon sessions. Being a divalent ion, Fe^{2+} also has potential to contribute to total hardness of water. Such a pattern is observed in the study area during pre-monsoon session, when iron shares a strong correlation with total hardness and calcium which is reflected in the pre-monsoon dendrogram as well, where iron is clustered with parameters defining the hardness character of groundwater.

Principal component and factor analysis

Principal component analysis (PCA) is a useful statistical tool for identifying patterns in a data set comprising of numerous parameters each with wide data range. Using principal component analysis method in factor analysis we can compress the data by reducing the number of dimensions without much loss of information. Factor analysis (FA) related to principal component analysis, generates an unobserved or latent variable which mainly compiles variations in three or more observed variables. Thus factor analysis as well as principal component analysis methods are put to use in water quality evaluation studies, where various parameters are tested in search for particular causes leading to change in quality of water (Helena et al. 2000; Reghunath et al. 2002; Yidana 2010). Prior to factor analysis on the study area data, the skewness and kurtosis of each parameter was calculated for post and pre-monsoon sessions. In the post-monsoon session, conductivity, total alkalinity, total hardness, bicarbonate and sulfate were found to be negatively skewed. All the other parameters were positively skewed. The kurtosis coefficients were greater than zero (95 % confidence) for all parameters except sodium and carbonate. In case of pre-monsoon session, all the analyzed parameters were positively skewed with kurtosis coefficients greater than zero (95 % confidence) except for magnesium and sodium. Principal component and factor analysis were performed for both sessions using the software StatistiXL following the Kaiser Normalization and Varimax Rotation procedures (Kaiser1958).

The Varimax rotated factor loadings for the post-monsoon session have been presented in Table 5. Six factors having Eigen values greater than one were generated in the post-monsoon session and rotations were completed in nine iterations. Figure 6a presents the scree plot generated based on the Eigen values of the components. Factor 1 explains 38.19 % of the total variance and besides fluoride and carbonate which have very low positive loadings, only pH contributes a significant factor loading which possibly indicates that and not the other ions analyzed do not contribute significantly towards pH of the groundwater. In that case, pH could possibly be controlled by biogenic factors (Helena et al. 2000). pH is negatively correlated to most of the analyzed parameters which is reflected by the negative factor loadings of the same. Factor 2 explains 16.73 % of the total

Table 5. Varimax rotated factor loadings for pre-monsoon session

Variable	Factor 1	Factor 2	Factor 3	Factor 4	Factor 5
pH	0.610	−0.015	0.272	−0.084	−0.107
TDS	−0.776	0.363	−0.454	−0.103	0.011
EC	−0.786	0.339	−0.442	−0.118	−0.007
TA	0.020	0.906	−0.173	0.319	−0.069
TH	−0.936	0.206	−0.056	−0.113	0.205
Ca^{2+}	−0.937	−0.108	−0.074	0.014	0.163
Mg^{2+}	−0.649	0.588	−0.014	−0.257	0.203
Na^{+}	−0.217	0.342	−0.782	0.247	−0.236
K^{+}	−0.069	0.183	−0.826	−0.021	0.161
Fe^{2+}	−0.816	−0.118	0.256	0.029	−0.206
CO_3^{2-}	0.047	0.102	0.102	0.914	−0.028
HCO_3^{-}	0.010	0.941	−0.211	0.103	−0.067
Cl^{-}	−0.923	−0.167	−0.203	−0.170	−0.028
SO_4^{2-}	−0.339	−0.096	−0.697	−0.169	0.568
PO_4^{3-}	−0.066	−0.050	−0.060	−0.068	0.951
F^{-}	0.098	0.145	−0.141	0.865	−0.088
Eigen value	6.518	3.176	1.886	1.355	1.027
% Variance	40.74	19.85	11.79	8.47	6.42
Cumulative % variance	40.74	60.59	72.38	80.85	87.26

variance and high positive loadings of total alkalinity, conductivity, sodium, potassium and carbonate are shown on this factor. High positive loading of potassium and carbonate and moderate loading value for sulfate demonstrates incidence of anthropogenic contamination sources. The study area is geologically composed of granite gneiss type rocks which are composed of alkali feldspars. Thus high positive loadings of sodium and potassium can also be attributed to dissolution of alkali feldspars. Factor 3 which explains 11.16 % of the total variance can be termed as the hardness factor as it shows high positive loadings of total hardness, calcium, magnesium and bicarbonate. High positive loading of alkalinity coupled with that of bicarbonate also indicates the salinity condition of the groundwater reserves. Moderately high scores of electrical conductivity and total hardness shown by factor 3 also imply chemical leaching of secondary minerals from the aquifer lithology (Routroy et al. 2013). Weathering of parent rock bodies and leaching of minerals leading to rock water interaction seem to be the factors controlling geochemistry of the groundwater. Factor 4 explains 7.52 % of the total variance and has very high positive factor loading for iron which can be attributed to the acidic character of the groundwater system. Factor 5 shows a high (>0.9) positive loading for phosphate and contributes 6.93 % to the total variance and can be associated to anthropogenic contamination factors. Factor 6 explains 6.26 % of the total variance and demonstrates moderate (>0.2) loadings for total hardness and calcium. Most of the other parameters have low or negative loadings thus indicating that this factor also partially defines the hardness quotient of groundwater.

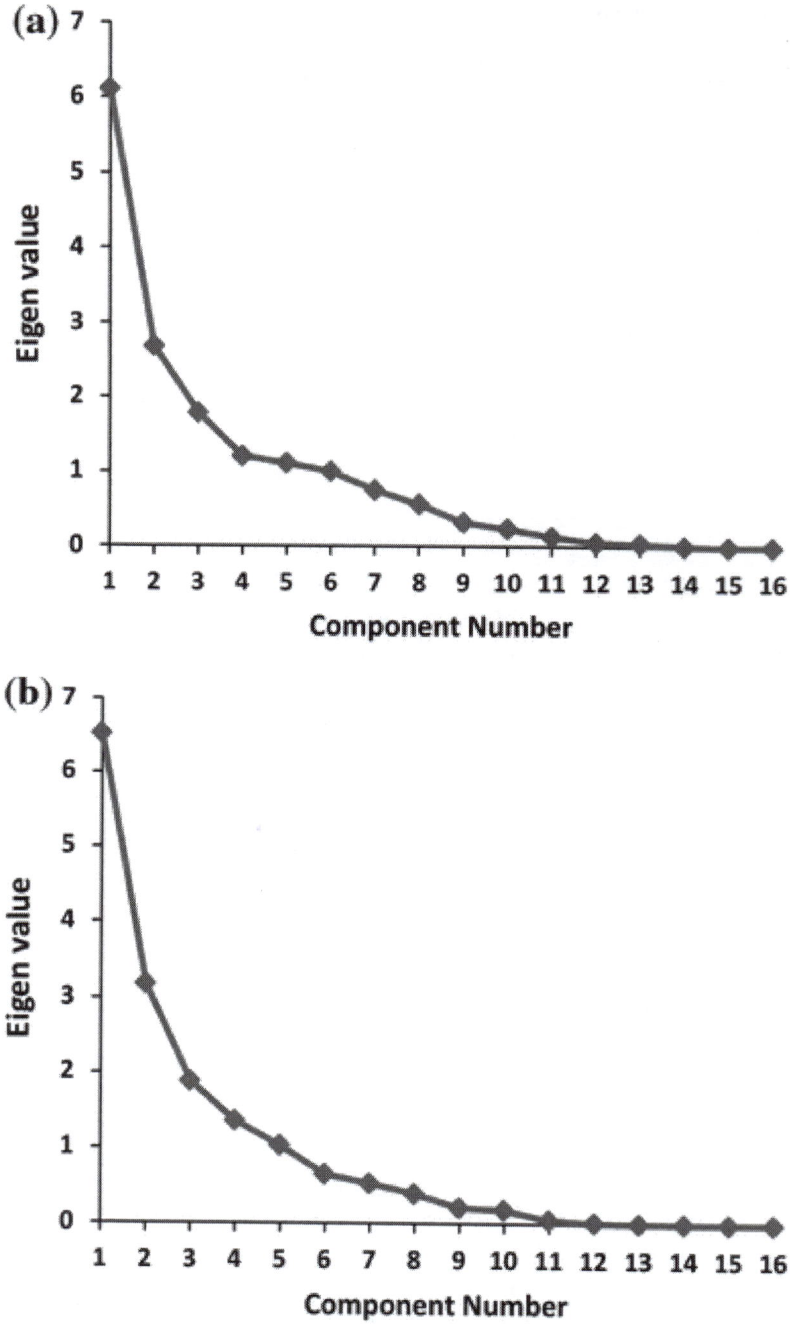

Figure. 6. **a** Scree plot generated through factor analysis for the post-monsoon session, **b** scree plot generated through factor analysis for the pre-monsoon session

The Varimax rotated factor loadings for the pre-monsoon session have been presented in Table 5. Five factors having Eigen values greater than one were generated in the pre-monsoon session and rotations were completed in seven iterations. Figure 6b presents the scree plot generated based on the Eigen values of the components. Factor 1 explains 40.74 % of the total variance. Similar to post-monsoon session, only pH has a high positive loading for factor 1. The correlationship of pH with rest of the analyzed parameters is negative or close to zero in the pre-monsoon session as well, which has been reflected through the loading values. Factor 2 explains 19.85 % of the total variance and shows high positive loadings for total alkalinity, magnesium and bicarbonate, an average loading value for sodium and a negative loading score for calcium. This factor can thus be assumed to demonstrate majorly the salinity character of groundwater. Factor 3 explains 11.79 % and shows low factor loadings for pH and iron. Based on the pH values at sampling locations of the study area, factor 3 can be assumed to define the acidic nature of groundwater. Factor 4 shows high loadings for carbonate and fluoride values, moderate factor loadings for alkalinity and it explains 8.47 % of the total variance. Fluoride shows a positive correlation with carbonate in the pre-monsoon session which is reflected in the factor loading values as well. The particularly strong interrelationship of fluoride with carbonate can be attributed to the calcite–fluorite geochemistry in groundwater where fluoride bears a negative correlation with calcium (as reported in the study) and a positive one with carbonate and bicarbonate. A moderate score for alkalinity and positive correlation of fluoride with alkalinity also demonstrates that unlike post-monsoon, rise in fluoride level in pre-monsoon might be associated with the alkalinity character of groundwater. Factor 5 explains 6.42 % of the total variance. Very high factor loadings are demonstrated by sulfate and phosphate in this case. A bulk of the study area is used for agricultural purposes. Thus factor 5 can be associated with anthropogenic contamination of groundwater occurring due to continuous use of fertilizers and pesticides in this area dominated by agricultural practices.

CONCLUSION

Excess fluoride has been observed in two locations of the study area, both in the post-monsoon and pre-monsoon sessions and the concentration of fluoride is seen to rise in the pre-monsoon session. Just like excess presence of excess fluoride in drinking can lead to dental and skeletal damages in the human body in form of fluorosis, similarly consuming water containing less than 0.6 mg/l fluoride can also lead to dental caries, especially in children in the age bracket of 6–10. In the present study, during both sampling sessions, groundwater in many locations have been found to contain less than 0.6 mg/l of fluoride thus making those areas prone to dental caries. Total alkalinity, the maximum permissible limit for which is 600 mg/l, is found to be in excess in 42 % of the samples in post-monsoon, whereas in pre-monsoon alkalinity is within desirable limit at all locations. Iron is present in excess at one location during post-monsoon, where during pre-monsoon iron content further rises up to more than 6 mg/l. Rest of parameters analyzed were all found to be present within the desirable limit in groundwater. Cluster analysis groups the parameters mainly into hardness, salinity, and dissolved solids factors in both sessions. Correlation analysis and Hierarchical cluster analysis results in post-monsoon do not reveal a particularly strong correlation or dependence of fluoride with any other parameter besides pH. In the pre-monsoon session fluoride is well clustered with the alkalinity parameters of water and seems to be regulated by them. The factor analysis results in pre-monsoon demonstrate strong affinity of fluoride for carbonate and also indicate anthropogenic factors playing a role in contaminating groundwater. Not all parameters follow the same trend in both sampling sessions regarding increase or decrease in their concentration in groundwater. The inter-relationships amongst them are also observed to change on a seasonal basis depending on recharge and discharge of groundwater reserves.

ACKNOWLEDGMENTS

The author, S. Das is thankful to UGC, New Delhi and Jadavpur University, for providing her the UGC Research Fellowship in Science for Meritorious Student 2012–2013. The other author (S.K. Nag) acknowledges the financial support received from the Centre of Advanced Study (CAS-Phase V), Department of Geological Sciences, Jadavpur University in conducting the field work. The authors gratefully acknowledge the support received from Dr. S. Gupta in analyzing the water quality parameters.

REFERENCES

1. Aghazadeh N, Mogaddam AA (2010) Assessment of groundwater quality and its suitability for drinking and agricultural uses in the Oshnavieh area, Northwest of Iran. J Environ Prot 1:30–40
2. APHA (American Public Health Association) (1995) Standard methods for examination of water and waste water. American Public Health Association, American Water Works Association and Water Pollution Control Federation, Washington
3. Bajpayee S, Das R, Ruj B, Adhikari K, Chatterjee PK (2012) Assessment by multivariate statistical analysis of ground water geochemical data of Bankura, India. Int J Environ Sci 3(2):870–880

4. Banoeng-Yakubo B, Yidana SM, Nti E (2009) Hydrochemical modeling of groundwater using multivariate statistical methods—the Volta Region, Ghana. KSCE J Civil Eng 13(1):55–63

5. Box GFP, Hunter WG, Hunter JS (1978) Statistics for experiments. An introduction to design data analysis and model building. Wiley, Toronto, p 653

6. Briz-Kishor BH, Murali G (1992) Factor analysis for revealing hydrochemical characteristics of a watershed. Environ Geol Water Sci 19(1):3–9

7. Chapman D (Ed) (1996) On the behalf of UNESCO, WHO. UXEP. Water quality assessments—a guide to use biota, sediments and water in environmental monitoring. F & F Spoil, London, Chapter 9

8. Cloutier V, Lefebvre R, Therrien R, Savard MM (2008) Multivariate statistical analysis of geochemical data as indicative of the hydrogeochemical evolution of groundwater in a sedimentary rock aquifer system. J Hydrol 353:294–313

9. Dart FJ (1974) The hazard of iron. Water and Pollution Control, Ottawa

10. Domenico PA, Schwartz FW (1990) Physical and chemical hydrogeology. Wiley, New York, p 824

11. Freeze RA, Cherry JA (1979) Groundwater. Prentice-Hall, Englewood Cliffs

12. Guler C, Thyne GD (2004) Hydrologic and geologic factors controlling surface and groundwater chemistry in Indian Wells-Owens Valley area, southeastern California, USA. J Hydrol 285:177–198

13. Güler C, Thyne G, McCray JE, Turner AK (2002) Evaluation and graphical and multivariate statistical methods for classification of water chemistry data. Hydrogeology 10:455–474

14. Gummadi S, Swarnalatha G, Vishnuvardhan Z, Harika D (2014) Statistical analysis of the groundwater samples from Bapatla Mandal, Guntur District, Andhra Pradesh, India. IOSR J Environ Sci Toxicol Food Technol 8(1):27–32

15. Gupta S, Banerjee S (2011) Fluoride in crops and dietary intake in a fluoride-endemic area of West Bengal. Research report, July-September 2011. Fluoride 44(3):153–157

16. Gupta S, Banerjee S, Saha R, Dutta JK, Mondal N (2006) Fluoride geochemistry of groundwater in Nalhati-1 Block of the Birbhum District, West Bengal, India. Res Rep Fluoride 39(4):318–320

17. Helena B, Pardo R, Vega M, Barrado E, Fernandez JM, Fernandez L (2000) Temporal evolution of groundwater composition in an alluvial aquifer (Pisuerga River, Spain) by principal component analysis. Water Res 34(3):807–816

18. Helsrup T, Jorgensen NO, Banoeng-Yakubo B (2007) Investigation of hydrochemical characteristics of groundwater from Cretaceous-Eocene limestone aquifers in southern Ghana and Togo using hierarchical cluster analysis. Hydrogeology 15:977–989

19. Hossien MT (2004) Hydrochemical evaluation of groundwater in the Blue Nile Basin, eastern Sudan, using conventional and multivariate techniques. Hydrogeol J 12:144–158

20. Jalali M (2006) Chemical characteristics of groundwater in parts of mountainous region, Alvand, Hamadan, Iran. Environ Geol 51:433–446

21. Jothivenkatachalam K, Nithya A, Chandra Mohan S (2010) Correlation analysis of drinking water quality in and around Perur block of Coimbatore District, Tamil Nadu, India. Rasayan J Chem 3(4):649–654

22. Kaiser HF (1958) The varimax criterion for analytic rotation in factor analysis. Psychometrika 23:187–200

23. Khazaei E, Stednick JD, Sanford WE, Warner JW (2006) Hydrochemical changes over time in the Zahedan aquifer, Iran. Environmental Monitoring and Assessment

24. Kortatsi BK (2007) Hydrochemical framework of groundwater in the Ankobra Basin, Ghana. Aquat Geochem 13(1):41–74

25. Kortatsi BK, Tay CK, Anornu G, Hayford E, Dartey G (2008) Hydrogeochemical evaluation of groundwater in the lower Offin basin, Ghana. Environ Geol 53:1651–1662

26. Kurumbein WC, Graybill FA (1965) An introduction to statistical models in geology. McGraw-Hill, New York

27. Machiwal D, Jha MK (2015) Identifying sources of groundwater contamination in a hard-rock aquifer system using multivariate statistical analyses and GIS-based geostatistical modeling techniques. J Hydrol Reg Stud.

28. Milovanovic M (2007) Water quality assessment and determination of pollution sources along the Axios/Vardar River, Southeast Europe. Desalination 213:159–173

29. Moghaddam A, Najib M (2006) Hydrogeologic characteristics of the alluvial tuff aquifer of northern Sahand Mountain slopes, Tabriz, Iran. Hydrogeol J 14:1319–1329

30. Moore CV (1973) Iron in: modern nutrition in health and disease. Philadelphia, Lea and Febiger, pp 297

31. Mrazovac S, Miloradov MV (2011) Correlation of main physicochemical parameters of some groundwater in northern Siberia. J Geochem Explor 108:176–182

32. Mukherjee B, Rao MG, Karunakaran C (1969) Genesis of kaolin deposits of Birbhum, West Bengal, India. Clay Miner 8:161–170

33. Nag SK (2014) Evaluation of hydrochemical parameters and quality assessment of the groundwater in Gangajalghati Block, Bankura District, West Bengal, India. Arabian J Sci Eng 39:5715–5727

34. Nag SK, Ghosh P (2013) Variation in groundwater levels and water quality in Chhatna Block, Bankura District, West Bengal, a GIS approach. J Geol Soc India 81(2):261–280

35. Nag SK, Ray S (2015) Hydrochemical evaluation of groundwater quality of Bankura I and II Blocks, Bankura District, West Bengal, India: emphasis on Irrigation and Domestic Utility. Arabian J Sci Eng 40:205–214

36. Nelson JD, Ward RD (1981) Statistical considerations and sampling techniques for ground—water quality monitoring. Groundwater 19(6):617–625

37. Pearson K (1896) Mathematical contributions to the theory of evolution, III. regression, heredity and panmixia. Philos Trans R Soc Lond 187:253–318

38. Ramesh Kumar A, Riyazuddin P (2008) Application of chemometric techniques in the assessment of groundwater pollution in a suburban area of Chennai city, India. Curr Sci 94(8):235–242

39. Reghunath R, Sreedhara Murthy TR, Raghavan BR (2002) The utility of multivariate statistical techniques in hydrogeochemical studies: an example from Karnataka, India. Water Res 36:2437–2442

40. Routroy S, Harichandran R, Mohanty JK, Panda CR (2013) A statistical appraisal to hydrogeochemistry of fluoride contaminated ground water in Nayagarh District, Odisha. J Geol Soc India 81:350–360

41. Sarbu C, Pop HF (2005) Principal component analysis versus fuzzy principal component analysis, a case study: the quality of Danube water (1985–1996). Talanta 6:1215–1220

42. Schiavo MA, Havser S, Gusimano G, Gatto L (2006) Geochemical characterization of groundwater and sub-marine discharge in the southeastern Sicily. Cont Shelf Res 26(7):826–834

43. Shrestha S, Kazama F (2006) Assessment of surface water quality using multivariate statistical techniques: a case study of the Fuji river basin, Japan. Environ Model Softw 22:464–475

44. Shrivastava VS, Patil PR (2002) Tapti river water pollution by industrial wastes: a statistical approach. Nat Environ Pollut Technol 1(3):279–283

45. Singh PK, Malik A, Sinha S (2005) Water quality assessment and apportionment of pollution sources of Gomti river (India) using multivariate statistical techniques—a case study. Anal Chim Acta 538:355–374

46. Sreedevi PD (2004) Groundwater quality of Pageru river basin, Cuddapah district, Andhra Pradesh, India. J Geol Sci 64:619–636

47. Subramani T, Elango L, Dhamodarasamy SR (2005) Groundwater quality and its suitability for drinking and agricultural use in Chithar River Basin, Tamilnadu, India. Environ Geo 47:1099–1110

48. Sunne Vazquez E, Sanchez Vila X, Carrera J (2005) Introductory review of specific factors influencing urban groundwater, an emerging branch of hydrogeology, with reference to Barcelona, Spain. Hydrogeol J 13:522–533

49. Vasanthavigar M, Srinivasamoorthy K, Vijayaragavan K, Rajiv Ganthi R, Chidambaram S, Sarama VS, Anandhan P, Manivannan R, Vasudevan S (2010) Application of water quality index for groundwater quality assessment: Thirumanimuttar Sub-Basin,Tamilnadu, India. Environ Monit Assess 171(1–4):595–609

50. WHO (2011) Guideline for drinking water quality, 4th edn. World Health Organization, Geneva

51. Yidana SM (2008) Groundwater resources management for productive uses in the Afram Plains area, Ghana. Doctoral Dissertation, Montclair State University, USA

52. Yidana SM (2010) Groundwater classification using multivariate statistical methods: Southern Ghana. J Afr Earth Sci 57:455–469

53. Yidana SM, Yidana A (2009) Assessing groundwater quality using water quality index and multivariate statistical analysis—the Voltaian basin, Ghana. J Environ Earth Sci.

54. Yidana SM, Ophori D, Banoeng-Yakubo B (2008a) Hydrogeological and hydrochemical characterization of the Voltaian Basin: the Afram Plains area, Ghana. Environ Geol 53:1213–1223

55. Yidana SM, Ophori D, Banoeng-Yakubo B (2008b) Hydrochemical evaluation of the Voltaian system—the Afram Plains area, Ghana. Environ Manage 88:697–707

56. Yidana SM, Ophori D, Banoeng-Yakubo B (2008c) Groundwater availability in the shallow aquifers of the southern Voltaian system: a simulation and chemical analysis. Environ Geol 55:1647–1657

57. Yidana SM, Ophori D, Banoeng-Yakubo B (2008d) Groundwater quality evaluation for productive uses—the Afram Plains area, Ghana. J Irrig Drain Eng 134(2):222–227

58. Zeng X, Rasmussen TC (2005) Multivariate statistical characterization of water quality in Lake Lanier, Georgia, USA. J Environ Qual 34:1980–1991

2

CHAPTER

Contributions of Multivariate Statistics in Oil and Gas Industry

Leandro Valim de Freitas[1,2], Ana Paula Barbosa Rodrigues de Freitas[2,3], Messias Borges Silva[2,3], Fernando Augusto Silva Marins[2], Estéfano Vizconde Veraszto[4,5], José Tarcísio Franco de Camargo[4] and J. Paulo Davim[6]

[1] *Petróleo Brasileiro SA (PETROBRAS), Brazil*

[2] *São Paulo State University (UNESP), Brazil*

[3] *University of São Paulo (USP), Brazil*

[4] *Municipal College Franco Montoro Professor (FMPFM), Brazil*

[5] *Campinas State University (UNICAMP), Brazil*

[6] *Aveiro University (UA), Portugal*

1. INTRODUCTION

This study aims to develop and validate multivariate mathematical models in order to monitor in real time the quality processing of derivatives in an oil refinery.

Methods heavily based on statistical and artificial intelligence as multivariate or chemometric methods have been widely used in the oil industry (KIM; LEE, KIM, 2009). Several articles have been written about applications of multivariate analysis to predict properties of oil derivatives (Santos Junior et al., 2005; Chung, 2007).

Pasadakis, Sourligas and Foteinopoulos (2006) have used the first six principal components of Principal Component Analysis (PCA) as input variables in nonlinear modeling of oil properties.

Pasquini and Bueno (2007) have proposed a new approach to predict the true boiling point of oil and its degree API (American Petroleum Institute) - a measure of the relative density of liquids by Partial Least Squares (PLS) and Artificial Neural Networks (ANN). Samples of mixtures oil were obtained from various producing regions of Brazil and abroad. In this application, the models obtained by the PLS method were superior to neural networks. The short time required for prediction the properties justifies the proposed of characterization the oil quicker to monitor refining processes.

Teixeira et al. (2008) in work with Brazilian gasoline used the multivariate algorithm Soft Independent Modeling of Class Analogy (SIMCA) for clusters analysis. Aiming to quantify the amount of adulteration of gasoline by other hydrocarbons, the PLS method was applied. Finally, the models were validated internally by cross-validation algorithm and externally with an independent set of samples.

Bao and Dai (2009) studied different multivariate methods, including linear and nonlinear techniques in order to minimize the error of prediction by models developed for quality control of gasoline. Lira et al. (2010) applied the PLS method for inference of the quality parameters: density, sulfur concentration and distillation temperatures of the mixture diesel / bio-diesel, providing great savings in time compared with the traditional methods by laboratory equipment.

Aleme, Corgozinho and Barbeira (2010) have conducted a study of classification of samples using the PCA method for discrimination of diesel oil type and the prediction of their origin.

Paiva Ferreira and Balestrassi (2007) have combined the Response Surface Method (RSM) of Design of Experiments (DOE) with Principal Component Analysis in optimizing multiple correlated responses in a manufacturing process.

Huang, Hsu and Liu (2009) have used Mahalanobis-Taguchi integrated with Artificial Neural Networks in data mining to look for patterns and modeling in manufacturing. Pal and Maiti (2010) have adopted the Mahalanobis-Taguchi algorithm to reduce the dimensionality of multivariate data and for optimization with Metaheuristics in the sequence.

Liu et al. (2007) have made inferences about quality parameters of jet fuel using Multiple Linear Regression (MLR) and ANN. The work showed that the performance of modeling by ANN was superior.

In optimization of multivariate models, there are applications combined with Multivariate Analysis of Metaheuristics, such as simulated annealing (SAUNIER, et al., 2009), genetic algorithm (GA) (Roy, Roy, 2009) tabu search (QI; SHI; KONG, 2010), particle swarm (Pal; Mait, 2010), and ant colony (Goodarzi; Freitas; Jensen, 2009; Allegrini; Oliveri, 2011).

With the objective of optimizing the dimensionality of multivariate models and avoid the overfitting phenomenon in determining principal components, Xu and Liang (2001) have used the Monte Carlo Simulation on simulated data sets and two real cases. Gourvénec et al. (2003) compared

Monte Carlo cross-validation with the traditional method of cross validation to determine the appropriate number of latent variables.

Adler e Yazhemsky (2010) have combined the Monte Carlo Simulation, PCA and Data Envelopment Analysis (DEA) in a context where there is a relatively large number of variables related to the number of observations for decision making. Llobet et al. (2005), by means a Multiple Criteria Decision-Making (MCDM) model, have used Fuzzy classification of samples of chips. For prediction oxidative and hydrolytic properties, was used an electronic nose based on PLS models, with prior selection of input variables by a GA Metaheuristic.

Wu, Feng and Wen (2011), in studies related to Botany, compared the performance of the growth of a tree species - Carya Cathayensis Sarg by PCA methods and Analytic Hierarchy Process (AHP), identifying the advantages and the disadvantages of each method, although the results obtained by both have been essentially identical.

Zhang et al. (2006) have combined the method Preference Ranking Organization Method for Enrichment Evaluations (PROMETHEE), from the Elimination et la Choix Traduisant Réalité (ELECTRE) and Geometrical Analysis for Interactive Assistance (GAIA) with PCA and PLS methods to classify 67 oils and determine an indicator of product quality. Purcell, O'Shea and Kokot (2007) also combined PROMETHEE and GAIA with PCA and PLS in studies related to cloning of sugarcane.

Regarding to the control charts designed to monitor the mean vector, Machado and Costa (2008) have studied the performance of T^2 charts based on principal components for monitoring multivariate processes. Lourenço et al. (2011) have used the principles of Process Analytical Technology (PAT) in the construction of control charts based on the scores of the first principal component versus time for the on-line monitoring of pharmaceutical processes.

Moreover, Multivariate Analysis is an important technique in various areas of knowledge such as Data Mining (Kettaneh; Berglund; Wold, 2005); Econometrics (Mackay, 2006); Marketing (Ahn; Choi; Han, 2007) and Supply Chain Management (Pozo et al., 2012).

2. APPLICATION: OIL REFINING

The first process in a refinery is atmospheric distillation or direct distillation, where components of crude oil are separated into different sections using different boiling points. The main products obtained in this process are: liquefied petroleum gas (LPG), naphtha - precursor of gasoline, jet fuel, diesel and fuel oil.

Additionally, refineries usually have a second tower, vacuum distillation, to produce diesel cuts. These intermediate streams feeding a chemical process called Fluid Catalytic Cracking (FCC). In this, two noble streams are generated: LPG, and gasoline. It is a refining scheme much more flexible, but though modern, may also present difficulties for framing products stricter specifications.

The production scheme level 3 is more flexible and cost effective than the previous one, because it uses the chemical process of Coking, which transforms a fraction of lower value - vacuum residue of distillation towers, in the noblest products like LPG, gasoline, naphtha and diesel oil.

This final refining scheme incorporates the process Hydrotreating of middle fractions generated in the Coker Unit, enabling increased supply of diesel with good quality. This scheme allows a more balanced supply of gasoline and diesel oil, producing more diesel and less gasoline than the previous settings.

Of course, there are other macro-processes and auxiliary processes such as water treatment plant, effluent disposal, sulfur recovery units, units of hydrogen generation and consequently other interconnections, details of which are not subject of this work (ANP, 2012).

3. METHODS

3.1. Acquisition Database: Infrared Radiation

In the oil industry, signs of infrared radiation generated by sensors are associated with the prediction of the quality of distillates such as naphtha, gasoline, diesel and jet fuel (Kim, Cho; Park, 2000).

Freitas et al. (2012) and Pasquini (2003) explain this instrumentation (Figure 1): the polychromatic radiation emitted by the source has a wavelength selected by a Michelson interferometer. The beam splitter has a refractive index such that approximately half of the radiation is directed to the fixed mirror and the other half is reflected, reaching the movable mirror and is therefore reflected by them. The optical path differences occur due the movement of the movable mirror that promotes wave interference.

An interferogram is obtained as a result of a graph of the signal intensity received by the detector versus the difference in optical path traveled by the beams. By calculating the Fourier Transform (FT) the interferogram can be written as a sum of sines and cosines (Tarumi et al, 2005) and in this case, happens to be called transmittance spectra, T (Forato; Filho; Colnago, 1997). Finally, the spectrum of transmittance, T, is converted to absorbance spectra, A, by co-logarithm of T (Suarez et al. May 2011). The absorbance can be interpreted as the amount of radiation that the sample absorbs and the transmittance, the fraction of radiation that the sample does not absorb. These phenomena occur depending on their chemical composition (Kramer; Small, 2007).

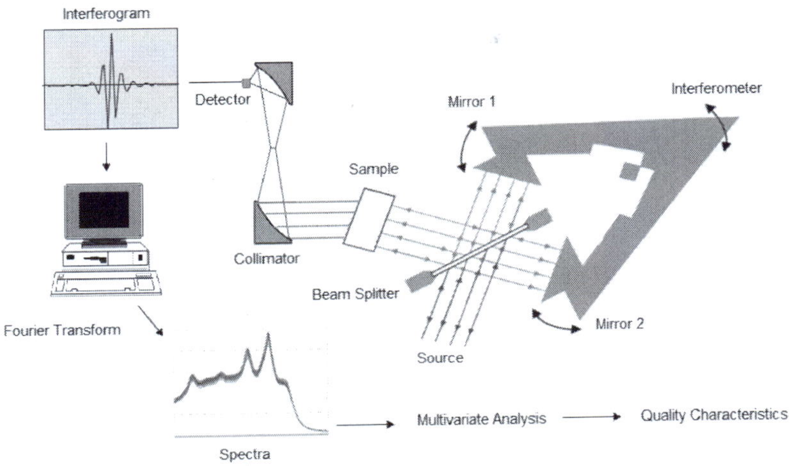

FIGURE 1. Scheme for technology acquisition database (Adapted from Pasquini, 2003**)**

The chemical bonds of the type carbon-hydrogen (CH), oxygen-hydrogen (OH) and nitrogen-hydrogen (NH), present in petroleum products (Pasquini; Bueno, 2007), are responsible for the absorption of infrared radiation, however, are not very intense and overlap. The broad spectral bands formed are difficult to interpret (Skoog; Holler; Crouch, 2007) due to the phenomenon of collinearity (Naes; Martens, 1984). The origin of this phenomenon is associated with the manner in which the infrared radiation interacts with matter and can be demonstrated by Quantum Mechanics at work Pasquini (2003).

These input variables (radiation absorbed), called X_i are correlated, so are said collinear or multicollinear (NAES et al., 2002). To illustrate the collinearity, X is a dummy matrix a_{ij} with i rows and j in terms columns, where a_{ij} is the radiation absorption of three samples i (i = 1, 2, 3) at two wavelengths j (j = 1, 2).

$$X = \begin{bmatrix} 1 & 4 \\ 2 & 5 \\ 3 & 6 \end{bmatrix} \quad (1)$$

The columns of X are linearly dependent, so the variables column j_1 and j_2 are colinear, that is, when increases j_1, j_2 increases proportionally. This causes the determinant of X'X to be zero, where X' is the transpose of matrix X.

$$X^T X = \begin{bmatrix} 14 & 28 \\ 28 & 56 \end{bmatrix} \quad (2)$$

Then, the det (X'X) = (14.56) - (28.28) = 0 and this according to Naes et. al (2002) means that there is a singular error matrix and that those erros are propagated when the dependent properties, Y, are determined by regression methods which are not based on the principal components, such as the MLR.

However, the multivariate approaches such as Principal Component Regression (PCR) and PLS have been quite appropriate due to dimensionality reduction, which creates a new set of variables called principal components (Rajalahti; Kvalheim, 2011). So with data mining for Multivariate Analysis, it is possible to relate the physicochemical properties (quality characteristics) of products with the chemical composition of the sample reflected by the absorption spectra. So once modeled a property, just a sample is subjected to infrared radiation to predict their properties.

3.1. Acquisition Data Base: Reference Properties

In this work were modeled properties of gasoline, diesel and jet fuel. For gasoline, the octane number and for diesel oil and jet fuel, the kinematic viscosity property.

According to Freitas (2012), kinematic viscosity of the diesel oil and jet fuel products is an important property in terms of its effect on power system and in fuel injection. Both high and low viscosities are undesirable since they can cause, among others, problems in fuel atomization. The formation of large and small droplets (low viscosity), can lead to a poor distribution of fuel and compromise the mixture air – fuel resulting in an incomplete combustion followed by loss power and greater fuel consumption.

The octane number of a gasoline is an important characteristic which is related to their ability to burn in spark-ignition engines. It is determined by comparing its tendency to detonate with the reference fuel with octane known under standard operating conditions.

When it comes to defining the octane required by engines, many countries use anti-knock index (I), defined by Equation 1:

$$I = \frac{MON + RON}{2} \quad (3)$$

where MON is the Motor Octane Number and RON is the Research Octane Number. The method MON measures the resistance to detonation when gasoline is being burned in the most demanding operating conditions and at higher rotations. The test is done in motors CFR (Cooperative Fuel Research), single-cylinder with variable compression ratio equipped with the necessary instrumentation in a stationary base, as shown in Figure 2.

Figure 2. CFR engine for MON octane (WAUKESHA, 2012)

The RON method evaluates the resistance of the gasoline to detonation under milder conditions and work in less rotation than that measured by octane number MON. The test is done in similar engines to those used for testing in MON octane.

It takes two hours and half to run the test MON and it is spent the same time for the test RON.

4. RESULTS AND DISCUSSION

Samples of gasoline, diesel and jet fuel, collected during 1 year, were subjected to laboratory tests, to determine the input variables, X_i, which are the infrared radiation absorbed, and the response variables, Y_i, that are physicochemical properties. The physicochemical properties will be predicted by PLS models.

The Table 1 summarizes the validation results of each model for products gasoline, diesel and jet fuel, where RMSEP (Root Mean Square Error of Prediction) corresponds to the standard deviation of the residuals (differences between measured and predicted values by the model).

The Figures 3-6 illustrate that the residues of models follow normal distribution, since in all cases the p-value was greater than 0.05.

Table 1. Summary of results of modeling and validation.

Product	Property	Number of samples	Latent Variables	RMSEP	Correlation
Diesel Oil	Viscosity (40ºC)	180	8	0.116 cSt	0.9368
Gasoline	MON	350	6	0.22	0.8723
Gasoline	RON	350	7	0.22	0.9891
Jet Fuel	Viscosity (-20ºC)	279	7	0.01 cSt	0.8836

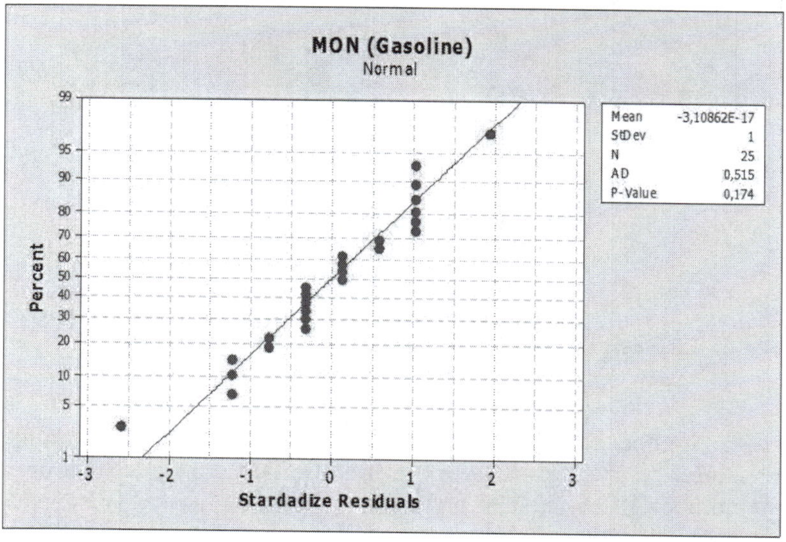

Figure 3. Normality Test for the property MON

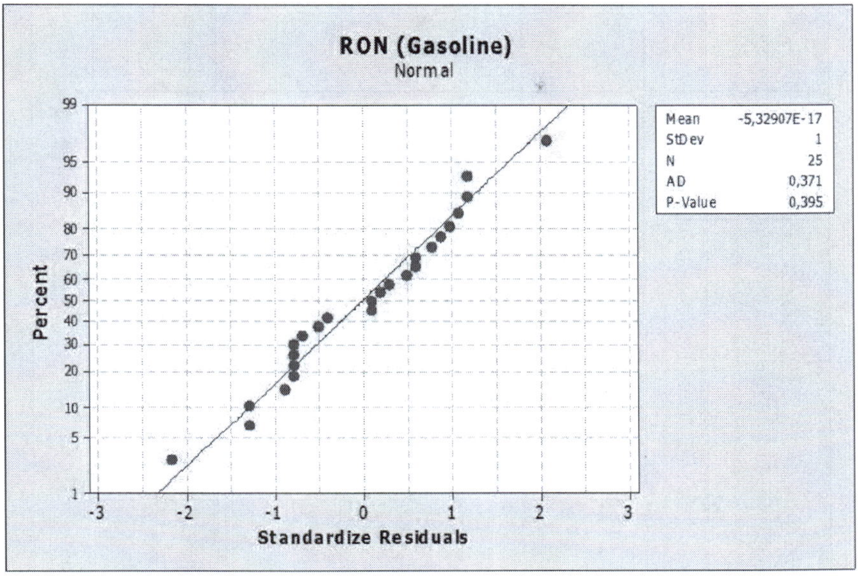

Figure 4. Normality test for the property ron

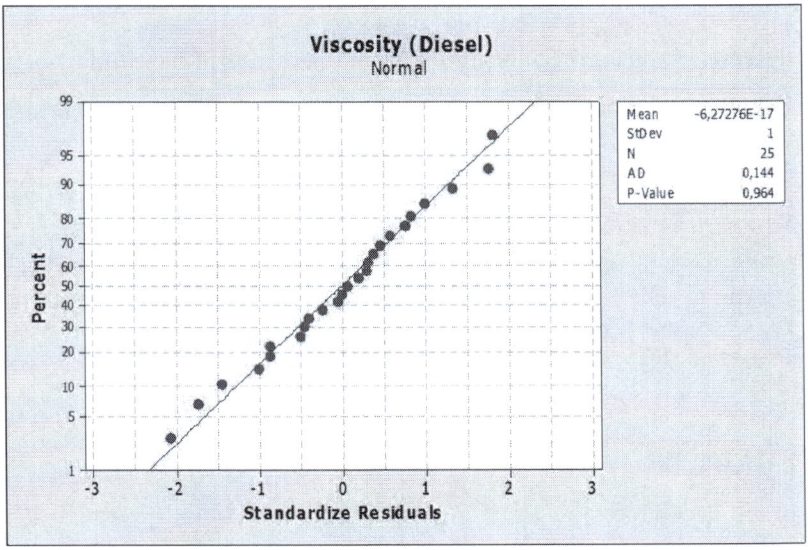

Figure 5. Normality Test for the property viscosity (diesel)

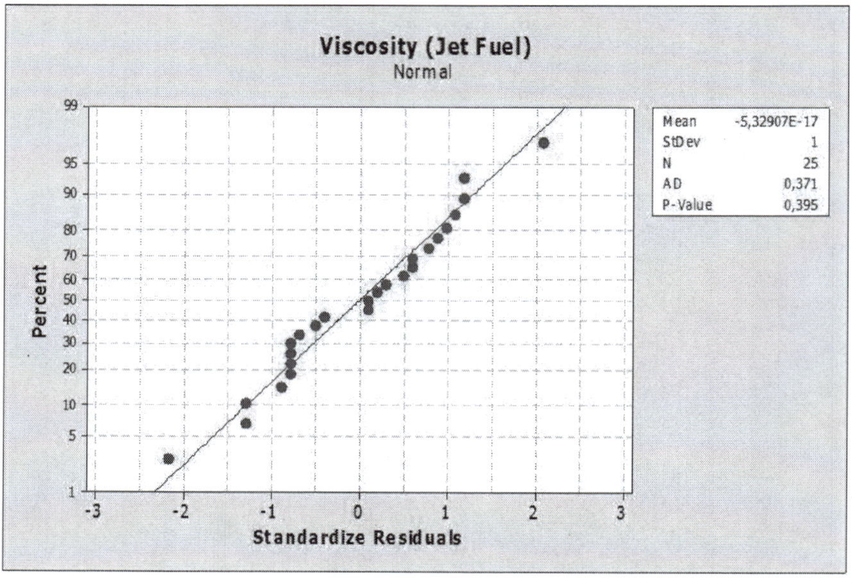

Figure 6. Normality Test for the property viscosity (jet fuel)

5. CONCLUSIONS

The following conclusions can be drawn from the results of this study:

It was possible to model mathematically the properties octane number and viscosity of the products gasoline, diesel and jet fuel.

The developed models were externally validated according to ASTM D-6122 and their predictions have precision equivalent to the reference methods.

The results were used in an oil refinery and contributed immensely to speed up the decision-making in blendings systems. Unlike the laboratory trials, the response time of a property along with the computational time does not exceed three minutes.

REFERENCES

1. Adler, N.; Yazhemsky, E. Improving discrimination in data envelopment analysis: PCA–DEA or variable reduction. European Journal of Operational Research, 2010, 202, 273-284.

2. AGENCY OF OIL, NATURAL GAS AND BIOFUELS. Schemes Production in Oil Refining:<http://www.anp.gov.br/?pg=7854&m=esquema+de+refi

no&t1=&t2=esquema+de+refino&t3=&t4=&ar=0&ps=1&cachebust =1331008874709>. Accessed in 06 March 2012.

3. Ahn, H.; Choi, E.; Han, I. Extracting underlying meaningful features and canceling noise using independent component analysis for direct marketing. Expert Systems with Applications, 2007, 33, 181-191.

4. Aleme, H. G.; Corgozinho, C. N. C.; Barbeira, P. J. S. Diesel oil discrimination by origin and type using physicochemical properties and multivariate analysis. Fuel, 2010, 89, 3151-3156.

5. Bao, X.; Dai, L. Partial least squares with outlier detection in spectral analysis: A tool to predict gasoline properties. Fuel, 2009, 88, 1216-122.

6. Forato, L. A.; Filho, R. B.; Colnago, L. A. Estudos de métodos de aumento de resolução de espectros de FTIR para análises de estruturas secundárias de proteínas. Química Nova, 1997, 20, 146-150.

7. Freitas, L. V.; Freitas, A. P. B. R.; Marins, F. A. S.; Loures, C. C. A.; Silva, M. B. Multivariate modeling in quality control of viscosity in fuel: an application in oil industry. Fuel Injection in Automotive Engineering. Rijeka: InTech, 2012.

8. Goodarzi, M.; Freitas, M.; Jensen, R. Ant colony optimization as a feature selection method in the QSAR modeling of anti-HIV-1 activities of 3-(3,5-dimethylbenzyl) uracil derivatives using MLR, PLS and SVM regressions. Chemometrics and Intelligent Laboratory Systems, 2009, 98, 123-129.

9. Gourvénec, S.; Pierna, J. A. F.; Massart, D. L.; Rutledge. An evaluation of the PoLiSh smoothed regression and the Monte Carlo Cross-Validation for the determination of the complexity of a PLS model. Chemometrics and Intelligent Laboratory Systems, 2003 68, 48-51.

10. Huang, C. L.; Hsu, T. S.; Liu, C. M. The Mahalanobis–Taguchi system – Neural network algorithm for data-mining. Expert Systems with Applications, 2009, 36, 5475-5480.

11. Kettaneh, N.; Berglund, A.; WOLD, S. PCA and PLS with very large data sets. Computational Statistics & Data Analysis, 2005, 48, 69-85.

12. Kim, D.; Lee, J.; Kim, J. Application of near infrared diffuse reflectance spectroscopy for on- line measurement of coal properties. Korean Journal of Chemical Engineering, 2009; 26; 489-495.

13. Kim, K.; Cho, I.; Park, J. Use of real-time NIR (near infrared) spectroscopy for the on-line optimization of a crude distillation unit. In: NPRA, Computer Conference. Chicago, 2000.

14. Kramer, K. E.; Small, G. W. Robust absorbance computations in the analysis of glucose by near-infrared spectroscopy. Vibrational Spectroscopy, 2007, 43, 440-446.

15. Lira, L. F. B.; Vasconcelos, F. V. C.; Pereira, C. F.; Paim, A. P. S.; Stragevitch, L.; Pimentel M.

16. F. Prediction of properties of diesel/biodiesel blends by infrared spectroscopy and multivariate calibration. Fuel, 2010, 89, 405-409.

17. Liu, H.; Yu, J.; Xu, J.; Fan, Y.; Bao, X. Identification of key oil refining technologies for China National Petroleum Co. (CNPC). Energy Police, 2007, 35, 2635 -2647.

18. Llobet, M. V. E.; Brezmes J.; Vilanova, X.; Correig, X. A fuzzy ARTMAP- and PLS-based MS e-nose for the qualitative and quantitative assessment of rancidity in crisps. Sensors and Actuators 2005, 106, 677-686.

19. Lourenço, V.; Herdling, T.; Reich, G.; Menezes, J. C.; Lochmann, D. Combining microwave resonance technology to multivariate data analysis as a novel PAT tool to improve process understanding in fluid bed granulation. European Journal of Pharmaceutics and Biopharmaceutics, 2011, 78, 513-521.

20. Machado, M. A. G.; Costa, A. F. B. The use of principal components and univariate charts to control multivariate processes. Operational Resarch, 2008, 28, 173-196.

21. Mackay, D. Chemometrics, econometrics, psychometrics - How best to handle hedonics?

22. Expert Systems with Applications, 2006, 17, 529-535.

23. Naes, T.; Isaksson, T.; Fearn, T.; Davies, T. Partial Least Squares. In: Multivariate calibration and classification. Chichester. NIR Publications, 2002.

24. Naes, T.; Martens, H. Multivariate calibration II. Chemometric methods. Trends in analytical chemistry, 1984, 3, 266-271.

25. Paiva, A. P. Metodologia de superfície de resposta e análise de componentes principais em otimização de processos de manufatura com múltiplas respostas correlacionadas. PhD Thesis; Itajubá Federal University, 2006.

26. Pal, A.; Maiti, J. Development of a hybrid methodology for dimensionality reduction in Mahalanobis–Taguchi system using Mahalanobis distance and binary particle swarm optimization. Expert Systems with Applications, 2010, 37, 1286-1293.

27. Pal, A.; Maiti, J. Development of a hybrid methodology for dimensionality reduction in Mahalanobis–Taguchi system using Mahalanobis distance and binary particle swarm optimization. Expert Systems with Applications, 2010, 37, 1286-1293.

28. Pasadakis, N.; Sourligas, S.; Foteinopoulos, Ch. Prediction of the distillation profile and cold properties of diesel fuels using mid-IR spectroscopy and neural networks. Fuel, 2006, 85, 1131-1137.

29. Pasquini, C. Near Infrared Spectroscopy: Fundamentals, Practical Aspects and Analytical Applications. Journal of Chemical Brazilian Society, 2003, 14, 198-219.

30. Pasquini, C.; Bueno, A. F. Characterization of petroleum using near-infrared spectroscopy: Quantitative modeling for the true boiling point curve and specific gravity. Fuel, 2007, 86, 1927-1934.

31. Pasquini, C.; Bueno, A. F. Characterization of petroleum using near-infrared spectroscopy: Quantitative modeling for the true boiling point curve and specific gravity. Fuel, 2007, 86, 1927-1934.

32. Pozo, C.; Fermenia, R. R.; Caballero, J.; Gosa, G. G. Jiménez, L. On the use of Principal Component Analysis for reducing the number of environmental objectives in multi- objective optimization: Application to the design of chemical supply chains. Chemical Engineering Science, 2012, 69, 146-158.

33. Purcell, D. E.; O'shea, M. G.; Kokot, S. Role of chemometrics for at-field application of NIR spectroscopy to predict sugarcane clonal performance. Chemometrics and Intelligent Laboratory Systems, 2007, 87, 113-124.

34. Qi, S.; Shi, W. M.; Kong, W. Modified tabu search approach for variable selection in quantitative structure–activity relationship studies of toxicity of aromatic compounds. Artificial Intelligence in Medicine, 2010, 49, 61-66.

35. Rajalahti, T.; Kvalheim, O. M. Multivariate data analysis in pharmaceutics: A tutorial review. International Journal of Pharmaceutics, 2011, 417, 280-290.

36. Roy, K.; Roy, P. P. Comparative chemometric modeling of cytochrome 3A4 inhibitory activity of structurally diverse

compounds using stepwise MLR, FA-MLR, PLS, GFA, G/PLS and ANN techniques. European Journal of Medicinal Chemistry, 2009, 44, 2913-2922.

37. Santos Jr V. O.; Oliveira F. C. C.; Lima, D. G.; Petry, A. C.; Garcia, E.; Suarez, P. A. Z.; Rubim,

38. J. C. A comparative study of diesel analysis by FTIR, FTNIR and FT-Raman spectroscopy using PLS and artificial neural network analysis. Analytica Chimica ACTA, 2005, 547, 188-196.

39. Saunier, O.; Bocquet, M.; Mathieu, A.; Isnard, O. Model reduction via principal component truncation for the optimal design of atmospheric monitoring networks. Atmospheric Environment, 2009, 43, 4940-4950.

40. Skoog, D. A.; Holler, F. J.; Crouch, S. R. Principles of Instrumental Analysis, 6 ed. Porto Alegre: Bookman, 2007.

41. Suarez, J. R. C.; Londoño, L. C. P.; Reyes, M. V.; Diem, M.; Tague, T. J.; Rivera, S. P. H. Fourier Transforms - New Analytical Approaches and FTIR Strategies: Open-Path FTIR Detection of Explosives on Metallic Surfaces. Rijeka: InTech, 2011. 520p.

42. Tarumi, T.; Small, G. W.; Combs, R. J.; Kroutil, R. T. Infinite impulse response filters for direct analysis of interferogram data from airborne passive Fourier transform infrared spectrometry. Vibrational Spectroscopy, 2005, 37, 39-52.

43. Teixeira, L. S. G.; Oliveira, F. S.; Santos, H. C. S.; Cordeiro, P. W. L.; Almeida, S. Q. Multivariate calibration in Fourier transform infrared spectrometry as a tool to detect adulterations in Brazilian gasoline. Fuel, 2008, 87, 346-352.

44. WAUKESHA. Industrial Engine - Industrial Gas Engine: <http://www.dresserwaukesha.com/index.cfm/go/list-products/productline/CFR-F1-F2- octane-category/>. Accessed in 06 March 2012.

45. Wu, D. S.; Feng, X.; Wen, Q.Q. The Research of Evaluation for Growth Suitability of Carya Cathayensis Sarg. Based on PCA and AHP. Procedia Engineering, 2011, 15, 1879-1883.

46. Xu, Q. S.; Liang, Y. Z. Monte Carlo cross validation. Chemometrics and Intelligent Laboratory Systems, 2001, 56, 1-11.

47. Zhang, G.; Ni, Y.; Churchill, J.; Kokot, S. Authentication of vegetable oils on the basis of their physico-chemical properties with the aid of chemometrics. Talanta, 2006, 70, 293-300.

3

CHAPTER

Identifying the Hydrochemical Characteristics of Rivers and Groundwater by Multivariate Statistical Analysis in the Sanjiang Plain, China

Yingjie Cao[1,2], Changyuan Tang[1,2], Xianfang Song[3], Changming Liu[3], Yinghua Zhang[3]

[1]*School of Environmental Science and Engineering, Sun Yat-Sen University, No. 135, Xingang Xi Road, Guangzhou 510275, People's Republic of China*

[2]*Graduate School of Horticulture, Chiba University, Matsudo 271-8510, Japan*

[3]*Key Laboratory of Water Cycle and Related Land Surface Process, Institute of Geographical Sciences and Natural Resources Research, Chinese Academy of Sciences, Datun Rd. A11, haoyang District, Beijing 100101, China*

ABSTRACT

Two multivariate statistical technologies, factor analysis (FA) and discriminant analysis (DA), are applied to study the river and groundwater hydrochemistry and its controlling processes in the Sanjiang Plain of the northeast China. Factor analysis identifies five factors which account for 79.65 % of the total variance in the dataset. Four factors bearing specific meanings as the river and groundwater hydrochemistry controlling processes are divided into two groups, the "natural hydrochemistry evolution" group and the "pollution" group. The "natural hydrochemistry evolution" group includes the salinity factor (factor 1) caused by rock weathering and the residence time factor (factor 2) reflecting the groundwater traveling time. The "pollution" group represents the groundwater quality deterioration due to geogenic pollution caused by elevated Fe and Mn (factor 3) and elevated nitrate (NO_3^-) introduced by human activities such as agriculture exploitations (factor 5). The hydrochemical difference and hydraulic connection among rivers (surface water, SW), shallow groundwater (SG) and deep groundwater (DG) group are evaluated by the factor scores obtained from FA and DA (Fisher's method). It is showed that the river water is characterized as low salinity and slight pollution, and the shallow groundwater has the highest salinity and severe pollution. The SW is well separated from SG and DG by Fisher's discriminant function, but the SG and DG can not be well separated showing their hydrochemical similarities, and emphasize hydraulic connections between SG and DG.

Keywords: Hydrochemistry Factor analysis Discriminant analysis The Sanjiang Plain

INTRODUCTION

The Sanjiang Plain, as a part of the Songhua River basin, is located in the northeast part of China, and it is one of the biggest and most important grain production bases in China. Large areas of wetlands have been exploited into crop lands since 1950s. In 2005, the area of crop lands was about 55,688 km^2, three times than that in 1954 (Song et al. 2008). On one hand, agriculture boom and development of cities have increased the demand of freshwater, on the other hand, the local surface water and groundwater quality have been deteriorated by the discharge of domestic and agricultural sewage. Elevated nitrate have been found both in rivers and shallow groundwater in the Sanjiang Plain (Cao et al. 2012). Another problem existing in the study area is the geogenic pollution caused by elevated Fe and Mn ranging from 0.3–24 to 0.2–0.4 mg/L, respectively (Yang et al. 2008), which have exceeded the drinking water sanitary standard (DWSS) of China (0.3 mg/L for Fe and 0.1 mg/L for Mn). Recent years, for the deterioration of shallow groundwater quality, well depths in some rural areas have reached from 50 m to almost 100 m. Extractions of deep groundwater may lead to pollutants transporting vertically from shallow groundwater to deep groundwater, resulting in a more serious problem to be dealt with. Therefore, it is necessary to identify the hydraulic connections among different water bodies such as rivers (SW), shallow groundwater (SG) and deep groundwater (DG).

Cao et al. (2012) have given a detailed analysis of the characteristics of nitrate in major rivers and aquifers of the Sanjiang Plain, but there are no reports found about the analysis of regional hydrochemical characteristics. Multivariate analysis, such as factor analysis (FA), discriminant analysis (DA) and cluster analysis (CA), aiming to interpret the governing processes through data reduction and classification, are recognized as powerful tools to deal with the increasing number of hydrochemical parameters. They have been effectively and widely applied to assessment of surface water quality (Panda et al. 2006; Shrestha and Kazama 2007; Yidana et al. 2008,2010), evaluation of the hydrogeochemical characteristic of groundwater (Cloutier et al. 2008; Papatheodorou et al. 2007; Yidana et al. 2010) and identification of groundwater contaminations (Dragon 2006; Kim et al. 2009; Zhou et al. 2007).

In this paper, two multivariate statistical analysis methods including FA and DA have been applied (1) to analyze the general characteristics of surface water (rivers) and groundwater hydrochemistry and their controlling factors; (2) to discriminate the hydrogeochemical difference among the SW, SG and DG; and (3) to deduce the hydraulic connectivity of these three water bodies based on hydrochemical similarities.

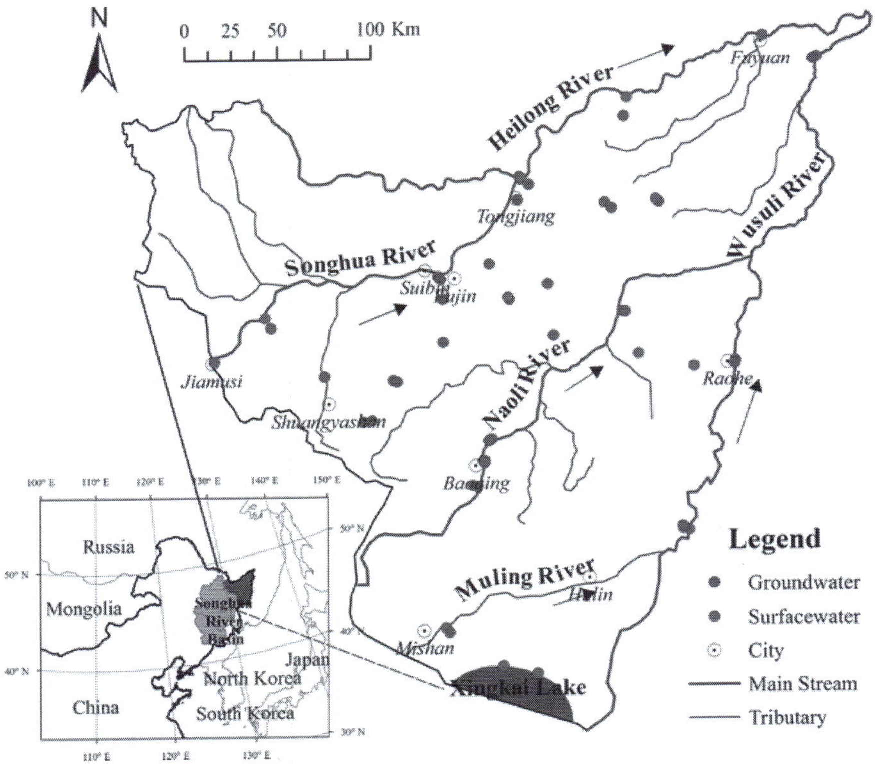

Fig. 1 Study area and sampling sites in the Sanjiang Plain

STUDY AREA

General descriptions

The Sanjiang Plain (129°46′42″–135°05′10″E, 46°06′30″N–48°28′06″N) is the alluvial plain within the Songhua River basin, northeast part of the Heilongjiang Province, China. The Xiao Xing Anling Mountains with a maximum elevation of 1,429 m above mean sea level and Wandashan Mountain with a maximum elevation of 831 m are found along the west side and southwest side of the study area. The north, east and southeast boundaries of the study area are the Heilong River, the Wusuli River and the Naoli River, respectively. The Songhua River and Naoli River flow through the plain and confluents with the Heilong River and Wusuli River, respectively (Fig. 1). The Sanjiang Plain has a generally flat topography declining from the southwest to the northeast with a slope gradient of about 1:5,000–1:10,000 except for some highlands in the center of the plain.

The study area located in temperate continental monsoon zone with annual mean temperature of 1.9 °C, the mean temperature in July and January is 22 and −21 °C, respectively (Sun and Song 2008). The annual precipitation is 500–650 mm, about 65 % of which occurring from June to August.

Geology and hydrogeology

The aquifers mainly consist of quaternary alluvial deposits, principally sandy gravels, medium and fine sands, sandy clays and silts. Underlain and surrounded by structurally disturbed base rocks of pre-Quaternary and older epochs, the thickness of aquifer varies from 50 to 300 m. Generally speaking, the thickness of sediments around the Suibin city in the west part of the study area is about 300 m and the size of deposits is coarser than other parts. The sediment around the line of Tongjiang-Fujin-Shuangyashan is mainly medium sands in a range of 50–100 m in thickness and the pre-Quaternary rocks crop out can be found beside the Fujin area (Yang et al. 2008). Vertically, the grain size of quaternary deposits in the layer above 60 m in depth is coarse, constituting the main aquifer in Sanjiang plain. Based on the ^{14}C (Yi and Wang2004) and CFCs (Cao et al. 2013) age dating results, the groundwater samples in this study have been classified into shallow groundwater and deep groundwater samples with sampling depth smaller than 50 m and larger than 50 m, respectively.

MATERIALS AND METHODOLOGY

Sample collection, chemical analysis and data preparation

34 groundwater water samples and 23 surface water samples were collected in the study area in September 2007 and September 2009 (Fig. 1). Field parameters pH, temperature (T), electrical conductivity (EC), redox potential (ORP) and dissolved oxygen (DO) of samples were measured by portable equipments (Horiba, Japan), and alkalinity was determined using the acid neutralizing titration on site. Samples were immediately filtered through a 0.45 μm filter and filled in two preclean 100 ml bottles at each site for major ion analysis. All samples were brought back to the laboratory and stored under a temperature of 4 °C. Cations such as K^+, Na^+, Ca^{2+} and Mg^{2+}, total Fe, Mn, Al, and dissolved silica (SiO_2) were analyzed by ICP-OES chromatography with the limit of detection (LOD) of 0.05 mg/L. Anions such as Cl^-, SO_4^- and NO_3^- were analyzed by ion chromatography (SHIMADZU LC-10AD) with the LOD of 0.01 mg/L. Charge balance error of the water samples is <5 % which is within the limits of acceptability.

Multivariate statistical analysis

Multivariate statistical analysis requires that the observations conform to normal distribution and the data transforms, such as log-transform (Cloutier et al. 2008), rank-ordered (Miller et al. 1997) and normalization (Ravichandrana et al. 1996; Ruiz et al.1990), are carried out before doing multivariate statistical analysis. In this study, the normality of variables was checked by one-sample *K–S* test. When data do not follow normal distribution, log-transform was selected to improve the data normality. ORP and Al have been removed from the multivariate statistical analysis, because the normality improvements of these two parameters were poor after the log-transformation. All the statistical analyses were conducted by the PASW Statistics 18.

Factor analysis

Factor analysis is a generic term used to describe a number of methods designed to analyze interrelationships within a set of variables or objects (Reyment and Joreskog1996). As a kind of dimension reduction technology in multivariate analysis, factor analysis can effectively reduce numerous data into few variables, the so-called factors. The underlying but unobservable or the latent processes which can not be directly revealed by the raw dataset could be determined through the use of factors and help to draw a more clear interpretation of the system. Factor analysis can be applied to the variables (R-mode) or to the cases (Q-mode). Here, main purpose is set to identify the hydrochemical characteristics and the

involved controlling factors. So the R-mode which is dealing with the variables analysis is selected.
The principle component method was applied to calculate the factor loading matrix. The number of factors were defined according to the criterion that only factors that account for variances >1 (eigenvalue-one criterion) should be included. The rationale for this criterion is that any component should account for more variance than any single variable in the standardized test score space (de Andrade et al. 2008). Varimax procedure (Carroll 1953) was used for the rotation of the retaining factors to yield an insight and informative interpretation or a more "simple structure" of the factor loading matrix. The terms 'strong', 'moderate', and 'weak' as applied to factor loadings refer to absolute loading values of ≥0.75, 0.75–0.50 and 0.50–0.30, respectively (Liu et al.2003). The factor scores were computed by the regression method which has the advantage of being the simplest technique and is easy to apply (Hardle and Simar2007). Dalton and Upchurch (1978) have shown that factor scores for each sample could be related to intensity of the process described by the factor. It is pointed out that extreme positive factor scores (≥1) reflected areas most affected and negative scores (≤−1) reflected areas unaffected by the process represented by the factor, while near-zero scores reflected areas affected to an average degree of the process.

In addition, the usefulness of factor scores could be summarized as two points (Reyment and Joreskog 1996). The first of these is that in general there will be fewer factors than original variables. Fewer maps or diagrams were needed for illustrating the spatial distribution of the variables or parameters and relationship between them. The second point is that for an orthogonal solution, the factor scores ("new" variables) will be uncorrelated, an important feature in using other statistical procedures, such as discriminant analysis and cluster analysis.

Discriminant analysis

Discriminant analysis is used in situations where the groups are known as a priori. The priori groups correspond to natural classifications or groups known from history or prior knowledge and also could have been formed by a cluster analysis performed on past data (Hardle and Simar 2007). In this study, the Fisher's discriminant method was used. Fisher's suggestion was to find the linear combination of datas which maximizes the ratio of the between-group-sum of squares to the within-group-sum of squares (Hardle and Simar 2007). The linear discriminant functions (LDF) is given by Eq. 1:

$$L = a'y = \sum_{j=1}^{p} a_j y_j \tag{1}$$

where L denotes LDF; p is the number of parameters used to classify a set of data into a given group; a_j is the weight coefficient assigned to a given selected parameter y_j.

Table 1. Descriptive statistics of physical parameters and chemical compositions for surface water and groundwater in the Sanjiang Plain

Group	Parameters	Unit	Mean	Median	Min	Max	STD
Surface water (SW)	EC	μs/cm	153.93	141.40	30.00	366.00	85.48
	pH		8.10	8.13	7.25	8.97	0.45
	T	°C	16.43	16.10	13.50	19.10	1.68
	ORP	mv	307.43	305.00	211.00	385.00	49.87
	DO	mg/L	6.17	6.30	0.00	9.65	1.79
	K^+	mg/L	1.45	1.24	0.00	6.55	1.43
	Na^+	mg/L	10.71	8.46	2.72	35.69	7.37
	Ca^{2+}	mg/L	21.95	21.91	9.49	43.60	8.66
	Mg^{2+}	mg/L	6.23	5.30	2.01	17.12	3.62
	HCO_3^-	mg/L	82.86	68.20	28.85	167.81	44.43
	Cl^-	mg/L	4.79	3.69	0.18	14.16	3.84
	SO_4^{2-}	mg/L	13.52	11.48	0.77	36.65	9.86
	NO_3^-	mg/L	2.44	1.51	0.00	8.48	2.24
	Fe	mg/L	2.42	2.13	0.40	5.17	1.48
	Mn	mg/L	0.32	0.12	0.03	1.61	0.42
	Al	mg/L	1.64	1.00	0.00	5.71	1.70
	SiO_2	mg/L	16.51	16.00	3.14	27.20	5.44
Shallow groundwater (SG ≤50 m)	EC	μs/cm	444.38	391.50	201.00	1042.00	245.71
	pH		7.44	7.38	6.30	8.45	0.49
	T	°C	9.60	8.50	6.30	15.40	2.68
	ORP	mv	186.58	168.00	-4.00	373.00	109.47
	DO	mg/L	2.64	2.32	0.27	5.80	1.64
	K^+	mg/L	3.26	1.63	0.00	11.20	3.62
	Na^+	mg/L	18.64	16.42	1.23	74.59	15.15
	Ca^{2+}	mg/L	55.39	44.02	6.69	187.28	44.62
	Mg^{2+}	mg/L	16.26	14.07	1.66	50.34	11.58
	HCO_3^-	mg/L	203.63	158.69	40.27	649.86	141.85
	Cl^-	mg/L	24.24	11.39	0.25	132.08	35.87
	SO_4^{2-}	mg/L	24.72	10.06	0.00	128.08	36.38
	NO_3^-	mg/L	19.07	2.72	0.00	153.29	35.78
	Fe	mg/L	4.23	2.54	0.00	24.59	5.49
	Mn	mg/L	0.52	0.35	0.00	2.02	0.59
	Al	mg/L	0.17	0.05	0.00	1.64	0.40
	SiO_2	mg/L	31.59	27.71	19.01	57.08	10.47
Deep groundwater (DG >50 m)	EC	μs/cm	313.05	298.00	151.50	536.00	123.56
	pH		7.45	7.49	6.89	8.08	0.39
	T	°C	9.70	9.40	6.80	16.80	2.71
	ORP	mv	171.55	149.00	28.00	348.00	123.27
	DO	mg/L	3.60	3.26	0.64	7.28	2.16
	K^+	mg/L	1.95	1.38	0.00	6.11	2.14
	Na^+	mg/L	15.59	14.67	3.38	27.04	6.31
	Ca^{2+}	mg/L	39.52	34.46	16.81	75.21	17.79
	Mg^{2+}	mg/L	10.63	11.77	3.70	20.12	4.69
	HCO_3^-	mg/L	177.93	186.23	83.90	238.69	47.39
	Cl^-	mg/L	7.91	2.74	1.02	54.09	15.51
	SO_4^{2-}	mg/L	8.52	3.50	0.00	34.45	10.87
	NO_3^-	mg/L	2.86	0.96	0.00	7.88	3.00
	Fe	mg/L	5.04	5.00	0.28	15.71	4.51
	Mn	mg/L	0.48	0.37	0.02	1.56	0.45
	Al	mg/L	0.53	0.00	0.00	5.71	1.72
	SiO_2	mg/L	33.08	37.40	4.28	52.77	15.20

RESULTS

Characteristics of general hydrochemistry

The statistical summaries of hydrochemical parameters for surface water (SW), shallow groundwater (SG) and deep groundwater (DG) in the

Sanjiang Plain are listed in Table 1. The average of pH is about 8.10 for surface water, which is more alkaline than that of shallow (7.44) and deep groundwater (7.45). The shallow groundwater has the highest EC with an average of 444 μs/cm, and the lowest EC is observed in the surface water with an average of 153 μs/cm.

Both surface water and groundwater are Ca-HCO_3 type. Prevalent cations are Ca^{2+} > Na^+ > Mg^{2+} > K^+. The average concentrations of Ca^{2+} are 21, 55 and 39 mg/L for SW, SG and DG, respectively. HCO_3^- is the dominate anion and is followed by $SO_4^{2-}$and Cl^-. Generally speaking, averages of the major ions were within acceptable ranges for domestic and irrigation uses. However, there are a few samples of elevated concentrations of NO_3^-, Fe and Mn suggesting local pollution in water environments. The max value of NO_3^- in the shallow groundwater is about 153 mg/L, three times the standard value (50 mg/L, DWSS). Average concentrations of Fe are 2.42 mg/L (SW), 4.23 mg/L (SG) and 5.04 mg/L (DG), respectively, which have exceeded the standard value of 0.30 mg/L in DWSS.

Factor analysis

The total dataset except for the parameters of Al and ORP was analyzed by R-mode factor analysis, and the result is showed in Table 2. Five factors which explain 79.65 % of the total variance are obtained based on the criteria for factor selection (eigenvalues >1.0). Communalities for the parameters are high (>0.70) for almost all variables except for Cl^- (0.60), SO_4^{2-} (0.68) and SiO_2 (0.66), and it indicates that the five-factor model explains the variability of almost all variables and can be used to indicate dominant processes controlling the local hydrochemistry without losing significant characteristics.

The factors are rotated by varimax method

Factor 1 explains 29.49 % of the total variance and has strong positive loadings on EC, Na^+, Ca^{2+}, Mg^{2+} and HCO_3^-, moderate positive loadings on Cl^- and K^+. Factor 2 explains 18.13 % of the total variance of the dataset and shows a significant bipolar characteristic. It has strong positive loadings on T and DO, and the negative pole has a strong loading on SiO_2. Factor 3 explains 12.57 % of the total variance of the dataset and shows strong positive loadings on Fe and Mn. Factor 4 explains 9.74 % of the total variance and has a strong positive loading on K^+. Factor 5 explains 9.73 % of the total variance and has a positive loading on NO_3^- and a moderate loading on SO_4^{2-}.

Table 2. Eigenvalues, percentage and cumulative percentage explained by each factor, factor loadings and the corresponding communalities. The factors with high loadings (> 0.75) are shown in bold

Factor	Eigenvalues	Percent of variance (%)	Cumulative percent of variance (%)
1	4.42	29.49	29.49
2	2.72	18.13	47.62
3	1.89	12.57	60.19
4	1.46	9.74	69.93
5	1.46	9.73	79.65

	Factor 1	Factor 2	Factor 3	Factor 4	Factor 5	Communalities
EC	**0.84**	−0.29	−0.16	0.20	0.09	0.86
pH	−0.19	0.66	0.10	−0.55	0.20	0.82
T	−0.33	**0.78**	−0.10	0.24	0.09	0.78
DO	−0.09	**0.78**	−0.19	−0.24	−0.11	0.72
K^+	0.22	0.11	0.19	**0.87**	0.04	0.85
Na^+	**0.79**	−0.10	0.29	−0.32	0.10	0.83
Ca^{2+}	**0.92**	−0.17	−0.06	0.09	−0.03	0.89
Mg^{2+}	**0.91**	−0.14	−0.08	0.13	0.12	0.88
HCO_3^-	**0.85**	−0.29	−0.06	0.05	−0.22	0.85
Cl^-	0.67	0.21	0.00	0.26	0.21	0.60
SO_4^{2-}	0.19	0.41	−0.20	−0.04	0.66	0.68
NO_3^-	0.01	−0.13	−0.11	0.02	**0.88**	0.80
Fe	−0.08	−0.07	**0.92**	0.07	−0.03	0.86
Mn	−0.01	−0.10	**0.88**	0.07	−0.25	0.86
SiO_2	0.13	**−0.77**	−0.01	−0.21	−0.04	0.66

Discriminant analysis

The Fisher's discriminant function was used to discriminate whether there is a significant hydrochemical difference and hydraulic connections between SW, SG and DG. Instead of the raw data set, here the factor scores obtained in the factor analysis were used as the input predictors. As mentioned above, the factor scores are orthogonal solutions so they are independent variables. This point is of benefit to carry out the discriminant analysis because colinearity of predictors will inflate the importance of some variables in their discriminant potentials. The number of Fisher's discriminant functions is related with the number of groups and number of predictors, and here two functions were identified and the function coefficients are displayed in Table 3.

Table 3. Discriminant function coefficients and the high coefficient values are shown in bold

Predictors	Fisher's method			
	Function 1	Function 2	Structure matrix	
Factor 1	**−0.84**	0.36	**−0.25**	0.40
Factor 2	**1.13**	0.17	**0.69**	0.38
Factor 3	0.12	**−0.46**	0.03	**−0.43**
Factor 5	−0.05	**0.79**	−0.01	**0.76**

DISCUSSION

Processes controlling the regional hydrochemistry

Based on the factor loadings, the five factors are divided into two groups, which represent two important geochemical processes controlling the regional hydrochemistry. The group 1 includes the factor 1 and factor 2, and the group 2 contains the factor 3 and factor 5. Factor 4 with strong loading on K^+ have no significant meaning and is eliminated from the following analysis. The chemical parameters EC, Na^+, Ca^{2+}, Mg^{2+} and HCO_3^- associated with factor 1 connote salinity enhancement caused by mineral dissolution (Yidana et al. 2008; Jang 2010), and is defined as "salinity" factor contributing to most of the TDS for water samples. It is also the most important process controlling the regional hydrochemistry due to its highest eigenvalue and ratio explaining the total variance.

In the cation Gibbs (1970) plot the water compositions are displayed as a function of the dominance of end-member components Na^+ at one end and Ca^{2+} at the other. The relative fractions for cations and anions are plotted against the total dissolved solids (TDS) on a logarithmic axis. Rock weathering and dissolution are apparently the dominant process influencing groundwater hydrochemistry in the study area (Fig. 2). But some of shallow groundwater showed a slight deviation from domain controlled by rock weathering and dissolution and branched into the seawater controlling domain which indicated that the shallow groundwater had been further salinized due to evaporation.

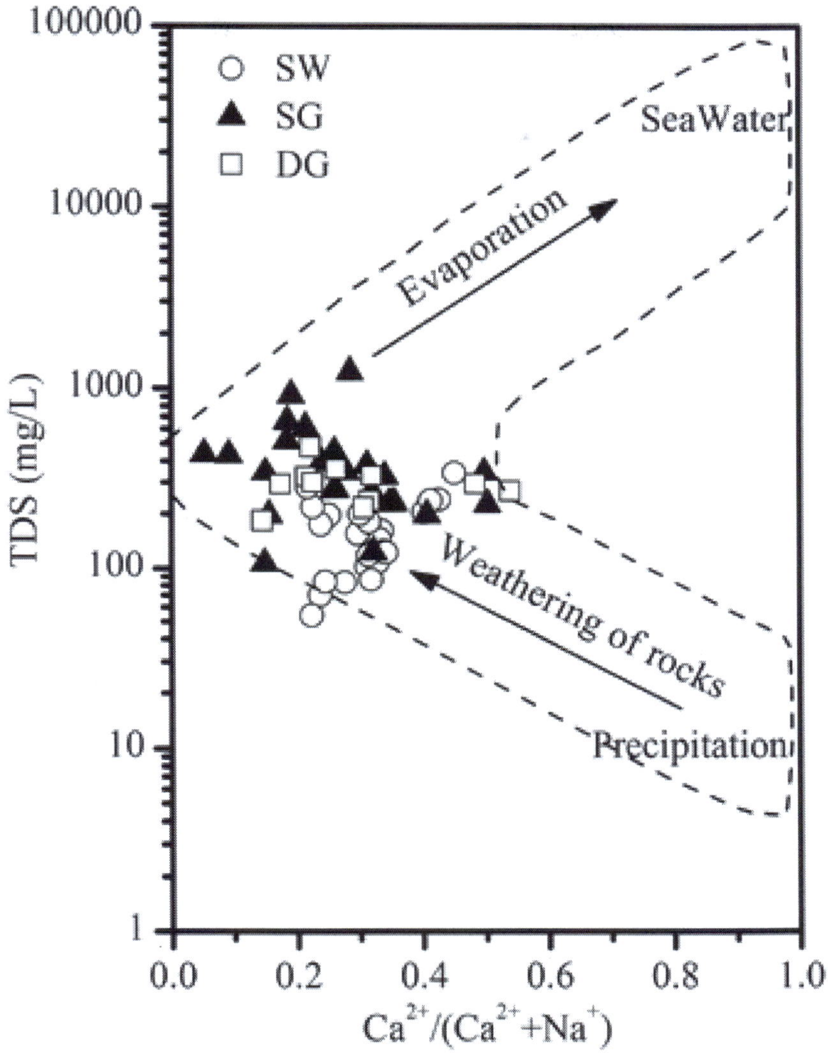

Figure 2. Gibbs (1970) plot for cations

Factor 2 has strong positive loadings on T and DO, and the negative pole has a strong loading on SiO_2. This factor is defined as "residence time" factor, which is reflecting the traveling time and the evolution of redox conditions for both surface water and groundwater. Yi (2001) reported that groundwater SiO_2 concentrations increase with sampling depths and the SiO_2 can be recognized as a residence time indicator in the Sanjiang Plain. The negative correlation between SiO_2 concentrations and DO (Fig. 3) implies that DO is consumed when the surface water recharge into the groundwater, and results in reducing environment for the groundwater with long residence time. The combination of factor 1 and 2 is recognized as the representative of natural hydrochemistry evolution processes (rock weathering and consumption of DO by organic matters) in the study area.

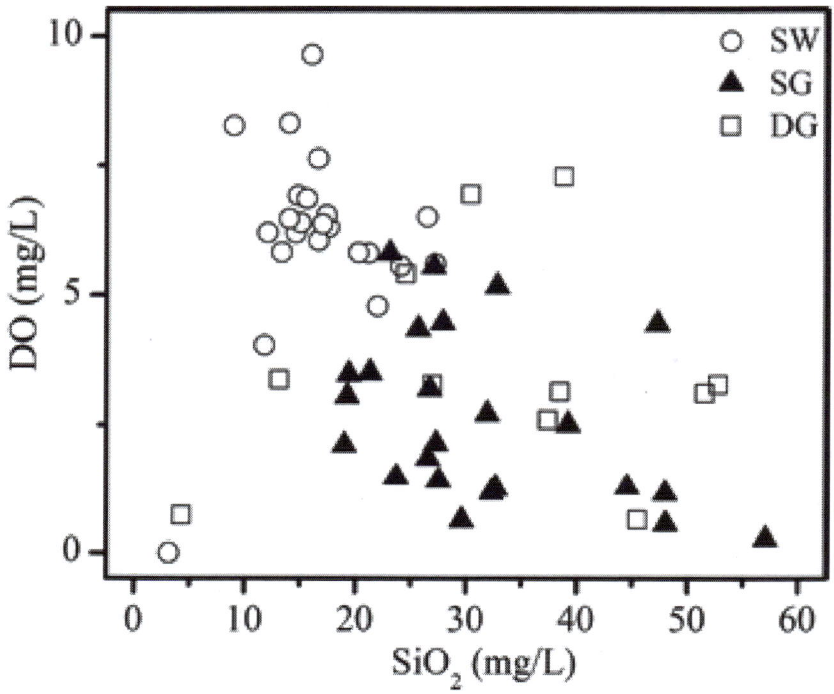

Figure 3. Relation between DO and SiO_2 concentrations

Factor 3 with high positive loadings on Fe and Mn is defined as "geogenic pollution" factor. Factor 5 with high positive loading on NO_3^- and a moderate loading on SO_4^{2-} is recognized as "anthropogenic pollution" factor, which is usually identified in the urban-, industry-, and agriculture-intensive areas due to the domestic wastewater discharge and agricultural activities (Dragon 2006; Lambrakis et al. 2004). The appearance of both NO_3^- and SO_4^{2-} in factor 5 implies the same sources for these two elements N and S. Major sources of NO_3^- come from fertilizer (chemical fertilizer and manure) and have been identified by Cao et al. (2012) by applying nitrogen isotope tracing tools. Papatheodorou et al. (2006) summarized three possible sources of SO_4^{2-} in groundwater and surface waters as atmospheric deposition, sulfate-bearing fertilizers and bacterial oxidation of sulfur compounds. In this study, considering that SO_4^{2-}appeared both in factor 5 with NO_3^-, it implies that SO_4^{2-} also presented a similar source with NO_3^-, coming from the use of fertilizers such as ammonium sulfate with major element of N and S. The combination of factor 3 and factor 5 is recognized as the "pollution" group which indicates major processes deteriorating the regional groundwater quality.

Hydrochemical differences

Hydrochemical differences have been detected among SW, SG and DG from Table 1. The natural hydrochemistry evolution processes and pollutions discussed in section "Processes controlling the regional hydrochemistry" are treated as two major reasons resulting in these differences. Here the factor scores are employed to quantify these differences. Surface water and groundwater samples are well separated by scores of the factor 1 and 2 (Fig. 4). Surface water is located in the left upper quadrant with factor 1 scores ≤0 and factor 2 scores ≥0. It implies that influence of rock weathering is lower in the surface water comparing to groundwater due to relative short residence time. In addition, contacting with air makes surface water have oxic conditions and high factor 2 scores.

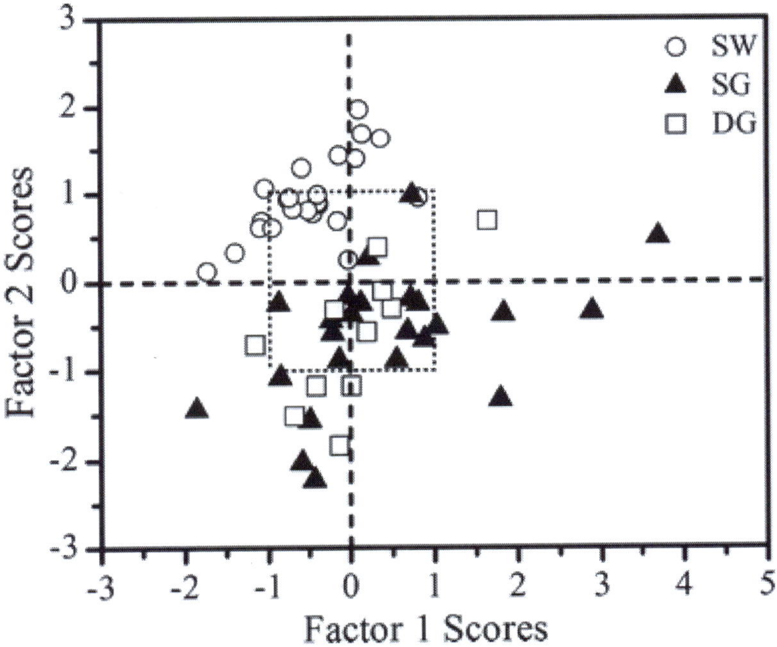

Figure 4. Factor scores between factor 1 and 2. The *dot line box* shows the range of factor scores between −1 and +1

Most of groundwater samples (SG and DG) are located in the lower half of Fig. 4 with factor 2 scores ≤0, implying that groundwater begins to enter into anoxic zones due to DO consumptions by organic matters without supply. The SG and DG mix together and are not well separated by factor 1 and 2, showing that their hydrochemistry are controlled by similar processes, here mainly referring to water–rock interactions. However, some "extremely affected" samples (mainly SG) with factor 1 score higher than +1 are characterized as high salinity because of the influence of evaporations, and some DG samples with factor 2 score lower than −1 implies enhance silicate weathering due to its long residence time.

Figure 5 presents the plot of factor scores for factor 3 and 5 to explain the hydrochemical differences caused by pollutions. Most of the SW and DG samples are confined by factor scores ranging from −1 to 1 showing that slight contaminations of surface water and deep groundwater. Specifically, the SW samples mainly located in the upper half of Fig. 5 with relative high factor 5 scores show that the major pollutant is NO_3^-, while for the DG, which are located in the lower half, have elevated Fe and Mn as the major pollutants. Comparing with SW and DG, SG is vulnerable to contaminations because groundwater table is shallow and the pollutants are easily passing through the unsaturated zones to recharge groundwater. Most of the SG samples locate out of the square bonded by the dot lines in Fig. 5 indicating serious pollutions caused by elevated NO_3^-, Fe and Mn.

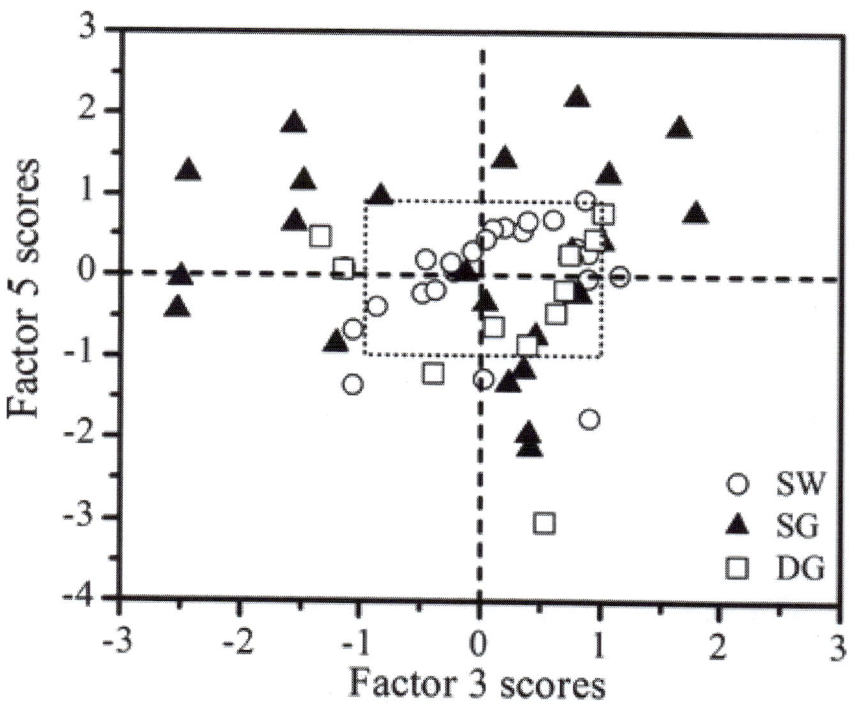

Figure 5. Factor scores between factor 3 and 5. The *dot line box* shows the range of factor scores between −1 and +1

Hydraulic connections revealed by hydrochemistry

In the Sanjiang Plain, many wells with depth deeper than 50 m have been dug for water supply due to the contamination of shallow groundwater. It is necessary to identify the hydraulic connections among SW, SG and DG, because extractions of deep groundwater may lead to pollutants transporting vertically from shallow groundwater to deep groundwater, and the contamination of deep groundwater is a more serious problem to be dealt with. In this study, the hydraulic connection is identified by evaluating the hydrochemistry similarity according to discriminant analysis.

The magnitudes of the coefficients given by Fisher's discriminant function directly show the importance of corresponding predictors and their contributions to discriminate the three groups (Table 3). For the discriminant function 1, the residence time factor (factor 2) with the largest coefficient 1.13 has the most substantial differentiating ability for discriminate scores and it is followed with the "salinity" factor (factor 2) which shows the second largest discriminate ability. Obviously, the function 1 discriminate the SW, SG and DG from the point view of natural hydrochemistry evolution. For the discriminant function 2, the "anthropogenic pollution" factor (factor 5) shows its largest coefficient of 0.79 and the "geologic pollution" factor (factor 3) has the second largest coefficient with a value of −0.43, and this function discriminates the three groups by the pollution indicators. Similar result is also confirmed by the structure matrix which shows the absolute correlation between each variable and any discriminant function.

The surface water (SW) is well separated with the groundwater samples (SG and DG) in the function 1 dimension, but for the SG and DG, samples gather together and cannot be well separated in both function 1 and function 2 dimension (Fig. 6). Specifically, the correct classification ratio for SW, SG and DG were 95.7, 54.2 and 63.6 %, respectively by applying the discriminant functions (Table 4). Moreover, 41.7 % of shallow groundwater samples are misclassified into deep groundwater group, while 36.4 % of deep groundwater samples are misclassified into shallow groundwater group. The poor discriminant ability for the SG and DG shows their hydrochemical similarities, and emphasized that tight hydraulic connections existed between SG and DG.

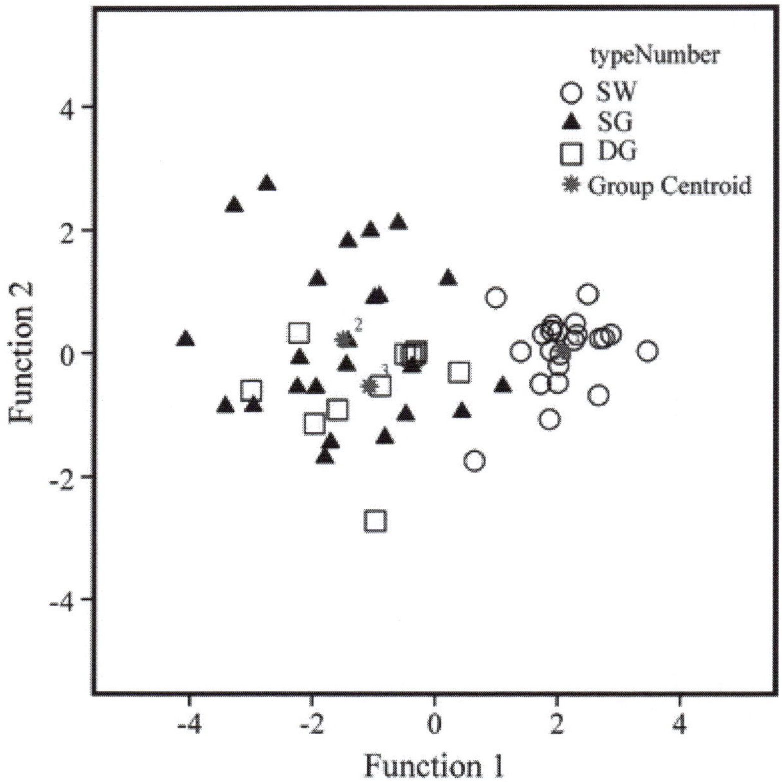

Figure 6. Separate groups by discriminant functions by Fisher's discriminant functions

Table 4. Classification results of DA

Origin groups	Predicted group membership			Total
	SW	SG	DG	
SW	22	0	1	23
SG	1	16	7	24
DG	0	2	9	11
SW	95.7 %	0.0 %	4.3 %	100 %
SG	4.2 %	54.2 %	41.7 %	100 %
DG	0.0 %	36.4 %	63.6 %	100 %

CONCLUSION

In the Sanjaing Plain, one of the most important grain production bases in China, water resources security has been recognized as a warranty of food security. Water pollution due to human activities threatens the regional water security, so in this paper, the major ion and pollutants hydrochemistry for both surface water and groundwater was studied by two multivariate analysis methods, factor analysis (FA) and discriminant analysis (DA), aiming to well understand the processes controlling the regional hydrochemistry, hydrochemical difference among SW, SG and DG, and hydrological connections for the three water bodies.

Factor analysis identified five factors accounting for 79.65 % of the total variance in the dataset. Two main factor groups were recognized as the main process controlling the regional hydrochemistry. The first factor group named "natural hydrochemistry evolution" consists of the "salinity" factor and "residence time" factor, and the second group revealing the "water pollution" is the combination of the "geogenic pollution" factor and "anthropogenic pollution" factor. Factor scores show the hydrological differences for SW, SG and DG. The surface water is characterized as having relative low salinity, slight contaminations and an oxygen-enriched state. The shallow groundwater has relative high TDS due to groundwater evaporation and is severely polluted with elevated nitrate, Fe and Mn. The deep groundwater has moderate TDS due to water–rock interaction during long residence time, and its major pollutants are Fe and Mn.

Fisher's discriminant method was used to discriminate the hydrochemical differences between SW, SG and DG, and shows that the factor group of "natural hydrochemistry evolution" had the most substantial discriminant ability for the surface water and groundwater. Poor discriminant ability for the shallow groundwater and deep groundwater reveals their hydrochemical similarity, indicating hydraulic connections between SG and DG.

ACKNOWLEDGMENTS

This work was funded by the Main Direction Program of Knowledge Innovation of Chinese Academy of Sciences: Study on water cycle, water resource carrying capacity and optimization configuration (No. KZCX2-YW-Q06-1). Bing Zhang, Peng Wang and Xin Liu (Chinese Academy of Sciences) provided valuable information and logistical assistance.

REFERENCES

1. Cao Y, Tang C, Song X, Liu C, Zhang Y (2012) Characteristics of nitrate in major rivers and aquifers of the Sanjiang Plain, China. J Environ Monit 14(10):2624–2633

2. Cao Y, Tang C, Song X, Liu C, Zhang Y (2013) Residence time as a key for comprehensive assessment of the relationship between changing land use and nitrates in regional groundwater systems. Environ Sci: Process Impacts 15(4):876–885

3. Carroll JB (1953) An analytical solution for approximating simple structure in factor analysis. Psychometrika 18(1):23–38

4. Cloutier V, Lefebvre R, Therrien R, Savard MM (2008) Multivariate statistical analysis of geochemical data as indicative of the hydrogeochemical evolution of groundwater in a sedimentary rock aquifer system. J Hydrol 353(3):294–313

5. Dalton MG, Upchurch SB (1978) Interpretation of hydrochemical facies by factor analysis. Ground Water 16(4):228–233

6. de Andrade EM, Palácio HAQ, Souza IH, de Oliveira Leão RA, Guerreiro MJ (2008) Land use effects in groundwater composition of an alluvial aquifer (Trussu River, Brazil) by multivariate techniques. Environ Res 106(2):170–177

7. Dragon K (2006) Application of factor analysis to study contamination of a semi-confined aquifer (Wielkopolska Buried Valley aquifer, Poland). J Hydrol 331(1):272–279

8. Gibbs RJ (1970) Mechanisms controlling world water chemistry. Science 170(3962):1088–1090

9. Hardle W, Simar L (2007) Applied multivariate statistical analysis. Springer Verlag, Berlin

10. Jang C-S (2010) Applying scores of multivariate statistical analyses to characterize relationships between hydrochemical properties and geological origins of springs in Taiwan. J Geochem Explor 105(1):11–18

11. Kim K-H, Yun S-T, Choi B-Y, Chae G-T, Joo Y, Kim K, Kim H-S (2009) Hydrochemical and multivariate statistical interpretations of spatial controls of nitrate concentrations in a shallow alluvial aquifer around oxbow lakes (Osong area, central Korea). J Contam Hydrol 107(3):114–127

12. Lambrakis N, Antonakos A, Panagopoulos G (2004) The use of multicomponent statistical analysis in hydrogeological environmental research. Water Res 38(7):1862–1872

13. Liu C-W, Lin K-H, Kuo Y-M (2003) Application of factor analysis in the assessment of groundwater quality in a blackfoot disease area in Taiwan. Sci Total Environ 313(1):77–89

14. Miller C, Denis J, Ator S, Brakebill J (1997) Nutrients in streams during baseflow in selected environmental settings of the Potomac River Basin. J Am Water Resour Assoc 33(6):1155–1171

15. Panda UC, Sundaray SK, Rath P, Nayak BB, Bhatta D (2006) Application of factor and cluster analysis for characterization of river and estuarine water systems–a case study: Mahanadi River (India). J Hydrol 331(3):434–445

16. Papatheodorou G, Demopoulou G, Lambrakis N (2006) A long-term study of temporal hydrochemical data in a shallow lake using multivariate statistical techniques. Ecol Model 193(3):759–776

17. Papatheodorou G, Lambrakis N, Panagopoulos G (2007) Application of multivariate statistical procedures to the hydrochemical study of a coastal aquifer: an example from Crete, Greece. Hydrol Process 21(11):1482–1495

18. Ravichandrana S, Ramanibai R, Pundarikanthan NV (1996) Ecoregions for describing water quality patterns in Tamiraparani basin, South India. J Hydrol 178(1–4):257–276

19. Reyment RA, Joreskog KG (1996) Applied factor analysis in the natural sciences. Cambridge Univ Press, Cambridge

20. Ruiz F, Gomis V, Blasco P (1990) Application of factor analysis to the hydrogeochemical study of a coastal aquifer. J Hydrol 119(1):169–177

21. Shrestha S, Kazama F (2007) Assessment of surface water quality using multivariate statistical techniques: a case study of the Fuji river basin, Japan. Environ Model Softw 22(4):464–475

22. Song K, Liu D, Wang Z, Zhang B, Jin C, Li F, Liu H (2008) Land use change in Sanjiang Plain and its driving forces analysis since 1954. Acta Geographica Sinica 63(1):93

23. Sun L, Song C (2008) Evapotranspiration from a freshwater marsh in the Sanjiang Plain, Northeast China. J Hydrol 352(1):202–210

24. Yang X, Yang W, Zhang F, Chu Y, Wang Y (eds) (2008) Investigation and assessment of groundwater resources potential and eco-environment geology in Sanjiang Plain. Geology Press of China

25. Yi X (2001) Metasilicic acid groundwater in the Sanjiang Plain. Ground Water 23(3):146–149

26. Yi X, Wang Z (2004) Shallow groundwater system in Sanjiang Plain. Ground Water 26(1):17–37

27. Yidana SM, Ophori D, Banoeng-Yakubo B (2008) A multivariate statistical analysis of surface water chemistry data: the Ankobra Basin, Ghana. J Environ Manage 86(1):80–87

28. Yidana SM, Banoeng-Yakubo B, Akabzaa TM (2010) Analysis of groundwater quality using multivariate and spatial analyses in the Keta Basin, Ghana. J Afr Earth Sci 58(2):220–234

29. Zhou F, Huang GH, Guo H, Zhang W, Hao Z (2007) Spatio-temporal patterns and source apportionment of coastal water pollution in eastern Hong Kong. Water Res 41(15):3429–3439

4

CHAPTER

Multivariate Analysis in Advanced Oxidation Process

Leandro Valim de Freitas[1, 2], Ana Paula Barbosa Rodrigues de Freitas[2, 3], Fernando Augusto Silva Marins[2], Carla Cristina Almeida Loures[2, 3], Messias Borges Silva[2, 3], Fatima Salman[3], Hilton Túlio Lima dos Santos[3], Gisella Lamas Samanamud[4]

[1] Petróleo Brasileiro SA (PETROBRAS), Brazil

[2] São Paulo State University (UNESP), Brazil

[3] University of São Paulo (USP), Brazil

[4] University of Texas at San Antonio (UTSA), USA

1. INTRODUCTION

Environmental degradation is attributable to improper industrial wastewater disposal, a situation that has caused serious contamination problems in many countries worldwide. Global consumption of potable water doubles every twenty years due to an exponential increase in world population (Han et al., 2009).

Industrial processes can create a wide variety of chemicals that pollute the air and water, with adverse impacts to ecosystems and humans. These impacts are caused by the polluting compounds that have toxic, carcinogenic, and also mutagenic properties (Busca et al., 2008). The treatment of wastewater containing phenolic compounds can be accomplished using applied principles of chemical oxidation, settling, membrane filtration, osmosis, ion, precipitation, and coagulation among other methods (Lin; Juang, 2009).

The treatment of hazardous wastes and reducing the presence of aqueous organic pollutants have placed focus on the use of alternatives to the standard environmental practices, such as the use of Advanced Oxidation

Processes (AOPs), especially for the wastewater treatment (Segura et al., 2009).

AOPs are considered a highly effective means of water treatment contributing to the effective removal of organic pollutants that, otherwise, are untreatable whether are adopted the traditional methods (Oller; Malato; Sanchez-Pérez, 2011).

The study of riverine water quality by countries has recently become problematic due to the progressive scarcity of resources (Ongley, 1998). The monitoring water quality and making-decisions based on qualitative data is a challenge for field researchers engaged in sample collection, storage, analysis, and the interpretations of results (Lermontov et al., 2008).

This study demonstrates the application of Multivariate Analysis (MA) in effluent treatment of polyester resin by using advanced oxidation processes (heterogeneous photocatalysis - UV/TiO_2). Exploring the relationship between methodologies and computational chemistry, the use of MA can modify the industrial processes and simplify experimental conditions, with a consequent improvement of processes, products, and the resolution of environmental issues.

2. POLYESTER RESIN

Polyesters and alkyd resins represent a class of polymers used in the manufacture of solvent-based paints due to the reduced cost and its versatility. These resins are the condensation products of polyols (e.g., glycerol, pentaerythritol), polybasic acids or their anhydrides (most phthalic anhydride) and monobasic fatty acids or oils. The term is typically restricted to polyester resins with acid or hydroxyl functional groups, which are relatively free of oil mixtures. The alkyd resins, a type of polyester resin, can be synthesized from renewable resources, i.e., vegetable oils as soybean oil, but most oils are coming from non-renewable sources (Abrafati, 1995; Weiss, 1997).

3. DYE INDUSTRY

Effluents from industries are highly complex dyes so they are not treated by conventional methods. Alternative matrices; AOPs; have been used to minimize the environmental impact of these industries on ecosystems (Nedhi; Sumner, 2003).

Industrial waste is classified into three main categories: wastewater, solid waste and air pollutants. Greater attention is given to wastewater since studies revealed content of significant amount of solvents (Metcalf; Eddy, 1991).

4. ADVANCED OXIDATION PROCESS

Advanced oxidation processes UV/TiO_2, UV/H_2O_2, UV/H_2O_2/Fe, O_3, O_3/Fe, O_3/TiO_2, UV-O_3/H_2O_2/Fe, are widely used in the degradation of effluents. These processes are characterized by the generation of free radicals in the organic matter degraded, and so are the polluting compounds mineralized or they are converted into a lower-chain or a less harmful process, in order to be subjected to a biological treatment subsequently (Thiruvenkatachari et al., 2007).

Oxidation processes are based on the generation of reactive species such as hydroxyl radical (OH). These radicals are highly reactive, non selective and may be used to degrade a wide range of organic pollutants. The OH radical is unstable and must be continuously generated *in situ*, by chemical or photochemical (Oliver; Hyunook; Pen-Chi, 2000). Table 1 shows reduction potential of various chemicals, and it can be observed that after fluorine, the hydroxyl radical is the one with higher oxidation potential (Domènech et al., 2001).

The treatment of hazardous waste and the presence of organic pollutants in water have increased the use of alternatives to environmental matrices such as the use of advanced oxidation processes (AOPs) in the treatment of wastewater (Segura et al., 2009).

According to Kusic, and Koppivanac Srsan (2007), the high standard reduction potential of the hydroxyl radical enables the oxidation of a wide variety of organic compounds to CO2, H2O and inorganic ions from heteroatoms.

According to Domenech et al. (2001), hydroxyl radicals can be produced by various advanced oxidation processes and heterogeneous and homogeneous systems, divided into two groups: non-photochemical and photochemical processes. Table 2 shows the procedures described above for the production of the hydroxyl radical.

4.1.... Advantages Of Advanced Oxidation Processes

AOPs have a number of advantages when compared to conventional oxidation processes (Gabardo Filho, 2005; Domènech et al., 2001):

Table 1. Reduction potential of some compounds

Compound	E^0 Reduction (V, 25 ºC)[1]
Fluorine(F_2)	3,03
Hydroxyl Radical ($\bullet OH$)	2,80
Atomic Oxygen (O_2)	2,42
Ozone(O_3)	2,07
Hydrogen peroxide (H_2O_2)	1,78
Radical Perhydroxyl ($HO_2\bullet$)	1,70
Chlorine dioxide	1,57
Hypochlorous acid (HCLO)	1,49
Chlorine (Cl_2)	1,36
Bromine (Br_2)	1,09
Iodine (I_2)	0,54

[1] Potential refers to the standard hydrogen electrode.
Source: Domenech et al. (2001)

Table 2. Exploited systems to produce hydroxyl radical

	With Irradiation	Without Irradiation
Homogeneous Systems	O_3/UV H_2O_2/UV H_2O_2/Fe^{2+}/UV	O_3/H_2O_2 O_3/OH^- H_2O_2/Fe^{2+}
Heterogeneous Systems	*Sc/O_3/UV *Sc/H_2O_2/UV *Sc/UV	Eletro-Fenton - -

*Sc: semiconductor (ZnO, TiO2, etc.)
Source: Morais (2005)

- Able to assimilate large variety of organic compounds;
- Full Mineralization of pollutants;
- Employed in the destruction of refractory compounds resistant to other treatments, such as the biologic;
- Can be integrated with other processes such as pre or post treatment;
- Used in high toxicity wastewater that can cause some difficulty in the treatment of biological process;
- Allow *in situ* treatment;
- Develop byproducts reaction intermediates that submitted to a post treatment may be mineralized;

- Improve organoleptic properties of treated water;
- Present high power with high oxidizing reaction kinetics.

4.2.... DISADVANTAGES OF ADVANCED OXIDATION PROCESSES

AOPs can be applied to certain types of waste under some restrictions, as follows (Domenech et al., 2001; Moral, 2005):

- Some processes are not available at appropriate scales;
- Costs can be high due to energy consumption;
- Some types do not apply to wastewater with high organic capability, turbidity, optical or color.

5. HETEROGENEOUS PHOTOCATALYSIS

The heterogeneous photocatalysis is a process based on the absorption of UV-visible by a solid semiconductor. In the interface area between the solution and the electrically excited solid, a degradation reaction or transformation of pollutants may occur, without changing the chemical structure of the semiconductor (Custo et al., 2006).

According to Nogueira and Jardim (1998), a semiconductor is characterized by:

- Valence bands, where vacancies are generated;
- Driving Bands that are generating electrons;
- A region between bands called bandgap;
- Semiconductor particles, when irradiated, absorb photons that may excite electrons from the valence band to the conduction band;

This generates vacancy electrons. The electron / vacancy pair migrates to the surface of the particle resulting in oxidation and reduction sites (Carp et al. 2004). Figure 1 illustrates schematically the behavior described above.

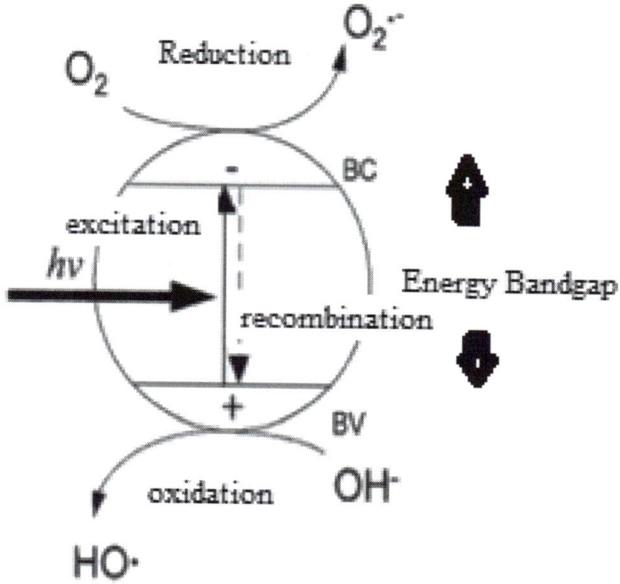

Figure 1. Electronic scheme of a photochemical process for heterogeneous photocatalysis (Ciola, 1981)

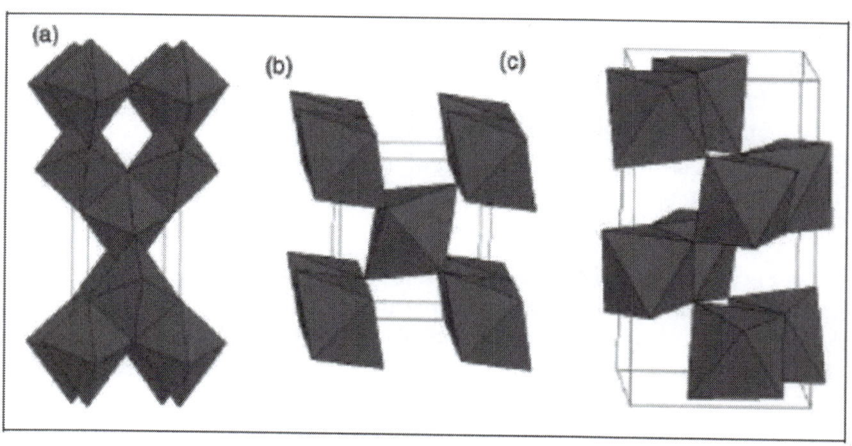

Figure 2. Crystal structures: anatase (a), rutile (b) and brookita (c)

TiO2 is a solid with a melting point of 1800 °C, and the ninth most abundant element corresponding to 0.63% of crust. The element described features 4 crystalline forms of anatase (tetragonal), brookita (orthorhombic), rutile (tetragonal) and TiO2 (B) monoclinic, as shown in Figure 2 (Carp et al., 2004).

6. DESIGN OF EXPERIMENTS

Design of Experiments (DOE) has been widely used in the optimization of process parameters and improving of quality products by the application of engineering concepts and Statistics (Wang, Huang, 2007).

Design of Experiments is defined as a set of statistical techniques applied to planning, conduction, analysis and interpretation of controlled tests, in order to find define factors that influence the values of a parameter or group of parameters (Bruns; Neto; Scarmínio, 2010).

According to Franceschini and Macchietto (2008), DOE is a statistical tool used to maximize the value of variable responses obtained on each experiment and also to minimize cost and time by reducing the number of experimental conditions.

Interactions between variables are considered in the experimental design and can be used for optimizing the operating parameters in multivariable systems (Ay; Catalkaya; Kargi, 2009).

According to Salazar (2009), the experimental has been studied as an important mathematical tool in the area of Advanced Oxidation Processes (Heterogeneous Photocatalysis). In this study, fractionated schedules for the degradation of organic matter and COD percentage of dairy effluent were used to obtain 93.70% of the treatment.

7. TAGUCHI ORTHOGONAL ARRAY

Taguchi method uses orthogonal arrays to study various factors with a small number of experiments (Sharma et al., 2005). Furthermore, the method provides other benefits, such as the reduction of process variability, low operating costs and expected results. (Barros; Bruns; Scarminio, 1995).

Rosa et al. (2009) define the Analysis of Variance (ANOVA) in the application of statistical analysis of Taguchi method in order to evaluate the significance of the parameters used in the process. A ANOVA table determines the most relevant parameters for the process according to equations 1, 2, 3 and 4:

-SS: Quadratic sum of the factors

$$SS = \sum_{i=1}^{n} (y_i - \bar{y})^2 \qquad (1)$$

- df: Degrees of freedom for each factor

$$df=N-1 \tag{2}$$

Where N means number of level for each factor

- MS: Mean square

$$MQ=\frac{SS}{df} \tag{3}$$

- F Test: Assessment of the significance of each factor

$$F = \frac{MS_{effect}}{MS_{error}} \tag{4}$$

8. MULTIVARIATE ANALYSIS

The Multivariate Analysis represents a set of statistical method in which most of variables of a data set are comprise information for decision-making (Rajalahti; Kvalheim, 2011), such as Multiple Linear Regression (MLR), Principal Component Analysis (PCA), Principal Component Regression (PCR) and Partial Least Squares Regression (PLS) (Otto, 2007).

The manufacturing processes can generally have correlated variables depending on the process quality that involves a large number of characteristics (Paiva, 2006).

The field of chemometrics is a multivariate analysis defined as the application of designs and mathematical and statistical methods to solve chemical problems. It is utilized to improve data collection and to allow extraction of useful and obtained data information (Hopke, 2003).

Paiva, Ferreira and Balestrassi (2007) combined DOE with Multivariate Analysis when optimizing multiple correlated responses in a manufacturing process.

8.1.... Multiple Linear Regression

The Multiple linear regression is a determining method of combinations of variables to achieve an optimal process or product. (Beebe; Pell; Seasholtz, 1998).

MLR property of interest relates to a linear combination of independent measurements. The modeling MRL can be represented by equation 5, where a set with n samples, i = 1 to n, Y is the response variable, X is the independent variable and i is the error estimation (Steiner et al., 2008)

$$y_i = \beta_0 + \sum_{j=1}^{k} \beta_i X_{ij} + e_i \tag{5}$$

According to Montgomery (2001), the method of Ordinary Least Squares (OLS - Ordinary Least Squares) to determine βi, minimize the sum of squared errors:

$$L = \sum_{i=0}^{n} \varepsilon_i^2 \tag{6}$$

$$= \sum_{i=1}^{n} \left(y_i - \beta_0 - \sum_{j=1}^{k} \beta_i x_{ji}\right)^2$$

Function L must be minimized in terms of $\beta_0, \beta_1, \ldots, \beta_k$.

$$\frac{\partial L}{\partial \beta_j} = -2 \sum_{i=1}^{n} \left(y_i - \beta_0 - \sum_{j=1}^{k} \hat{\beta}_i x_{ji}\right) \tag{7}$$

Simply stated, there is equality:

$$\hat{\beta}_0 \sum_{i=1}^{n} x_{ik} + \hat{\beta}_1 \sum_{i=1}^{n} x_{ik} x_{i1} + \hat{\beta}_2 \sum_{i=1}^{n} x_{ik} x_{i2} + \ldots$$

$$+ \hat{\beta}_k \sum_{i=1}^{n} {x_{ik}}^2 = \sum_{i=1}^{n} x_{ik}\, y_i$$

In matrix notation, there is Y = βX + ε, where:

$$y = \begin{bmatrix} y_1 \\ y_2 \\ \vdots \\ y_n \end{bmatrix}, X = \begin{bmatrix} 1 & x_{11} & x_{11} & \cdots & x_{1k} \\ 1 & x_{21} & x_{22} & \cdots & x_{2k} \\ \vdots & \vdots & \vdots & \vdots & \vdots \\ 1 & x_{n1} & x_{n2} & \cdots & x_{nk} \end{bmatrix}, \beta = \begin{bmatrix} \beta_1 \\ \beta_2 \\ \vdots \\ \beta_k \end{bmatrix} \text{ e } \varepsilon = \begin{bmatrix} \varepsilon_1 \\ \varepsilon_2 \\ \vdots \\ \varepsilon_n \end{bmatrix}$$

Then:

$$L = \sum_{i=0}^{n} \varepsilon_i^2 = \varepsilon^T \varepsilon = (y - X\beta)^T (y - X\beta)$$

$$L = y^T y - \beta^T X^T y - y^T X \beta + \beta^T X^T X \beta$$

$$L = y^T y - 2\beta^T X^T y + \beta^T X^T X \beta$$

Minimizing the function:

$$\frac{\partial L}{\partial \beta} = -2X^T y + 2X^T X \hat{\beta} = 0$$

Hence, is obtained:

$$X^T X \hat{\beta} = X^T y$$

and consequently, the coefficients are determined by Equation 6:

$$\hat{\beta} = (X^T X)^{-1} X^T y$$

9. MATERIALS AND METHODS

This work was performed at the Laboratory of the Environmental Engineering Department of the Chemical Engineering of School of Lorena EEL-USP. Polyester resin effluent was supplied by Valspar industry, located in São Bernardo do Campo, State of São Paulo. Statistical analyses were performed by Statistica version 2.0, available at the College.

Samples were stored cold chamber at EEL-USP at 4ºC. The effluent from the oxidation reaction was conducted into a Germetec tubular reactor, model FPG-463/1, with a nominal volume of approximately 1 L, receiving irradiation from a GPH-463T5L mercury lamp of low pressure, which emits a UV radiation at 254 nm with a power of 15 W and 21 W and protected by a quartz tube. The manufactured reactor model is shown in Figure 3.

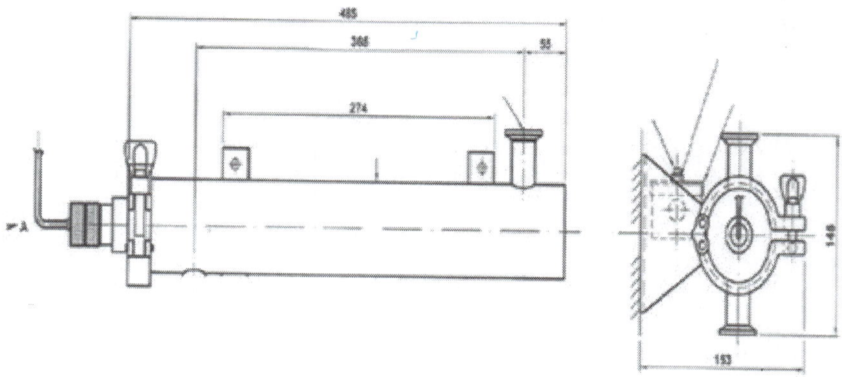

Figure 3. Tubular reactor used for photochemical treatment

Experimental design Taguchi L_9 with advanced oxidation process and heterogeneous photocatalysis were used for 1.0 liters of fresh effluent and 2 liters of distilled water, previously homogenized and conditioned at room temperature. Semiconductor titanium dioxide (TiO2), and the amount of H2O2 (30% w / v) were added during the initial 50 minutes of 1-hour total reaction, using burettes of 25 and 50 ml. The temperature of the medium reaction during the whole period of the photocatalytic process was controlled at 25 °C by using an Ophterm DC1 thermostatic bath. pH reaction was performed using a combined glass electrode adapted to the shell. This was connected to the potentiostat digital Digimed. A centrifugal pump was used for conducting the effluent from the tubular reactor to the storage tank. Ultraviolet lamps of 15 and 21 watts were used. Figure 4 presents the detailed scheme for treatment with AOP showing the experimental procedure.

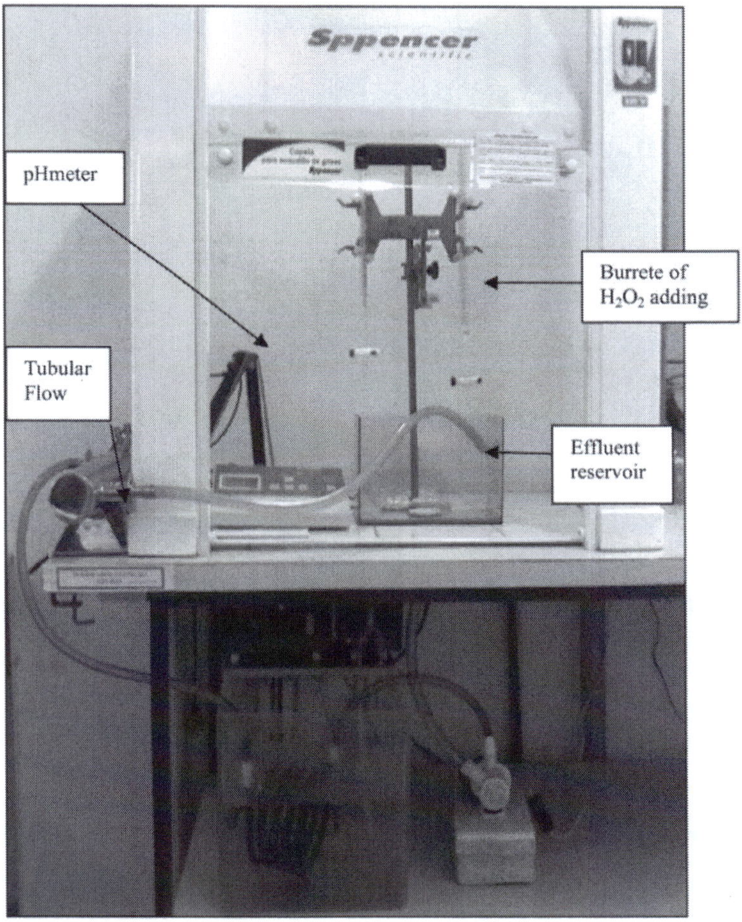

Figure 4. Layout of the process of photochemical treatment

10. RESULTS AND DISCUSSION

TOC (Total Organic Carbon) analysis of the effluent was conducted according to relevant physical-chemical aspects; that is, parameters monitored by environmental agencies (Morais, 2005). Figure 5illustrates the appearance of the effluent polyester resin *in natura.*

Figure 5. Effluent polyester resin in natura

The statistical planning performed is represented by Taguchi L9 orthogonal array to which the response variable was TOC and independent variables as factors proposed for this stage were: pH, titanium dioxide, hydrogen peroxide and UV radiation power. Table 3 shows the variables with treatment of levels with selected AOP.

Table 3. Control variables and their levels

Control variables (factors)	Level 1	Level 2	Level 3
A- pH	3,0	5,0	7,0
B- TiO2 [g/L]	0,083	0,167	0,250
C- H2O2*[g]	120,0	151,0	182,0
D- UV [W]	Sem	15	21

* $[H_2O_2]$ = 30 % m/m

Initially, the mass of H_2O_2 (30% w / w) is calculated by a stoichiometric ratio that depended on the organic load of the effluent. This obtained a mass of H2O2 of 50 g per liter of effluent.

The amount of TOC of the effluent *in natura* had a mean value of 7920mg / L that was subjected to pre-treatment. For each experiment amount of TOC, a sample in a 60-minute reaction was determined. The percent reduction of TOC is calculated by equation 7.

$$\%\text{ reduction of TOC} = \frac{TOC_{in\ natura} - TOC_{t=60\ min}}{TOC_{in\ natura}} \tag{7}$$

Table 4 shows the arrangement of orthogonal Taguchi L9 for the treatment of effluent polyester resin using AOP.

Table 4. L_9 Taguchi Orthogonal Array with 4 factors and 3 levels each

Experiment	pH Factor A	TiO2 Factor B	H_2O_2 Factor C	UV Factor D
1	1	1	1	1
2	1	2	2	2
3	1	3	3	3
4	2	1	2	3
5	2	2	3	1
6	2	3	1	2
7	3	1	3	2
8	3	2	1	3
9	3	3	2	1

Table 5 shows the percentage change in COT response to experiments using experimental design L9. Experiments 3 and 4 had a higher percentage of TOC removal for the advanced oxidation process (Heterogeneous Photocatalysis).

Table 5. Results of the first replica of the percentage reduction obtained in experiments for an initial TOC of 7920mg / l.

Experiment	pH Factor A	TiO2 Factor B	H2O2 Factor C	UV Factor D	Replica 1: reduction of total organic carbon (%)
1	1	1	1	1	31,970
2	1	2	2	2	34,981
3	1	3	3	3	39,489
4	2	1	2	3	37,216
5	2	2	3	1	29,962
6	2	3	1	2	30,095
7	3	1	3	2	30,549
8	3	2	1	3	33,504
9	3	3	2	1	28,182

Table 6 shows a replica of the experiment and the experimental conditions 3 and 4 that achieved significant values again.

Statistical analysis of Taguchi L9, on Figure 6, showed the most significant parameters for the degradation of organic matter in the wastewater, the latter reflecting of pH = 3, adjusted at a low level and factors set at a maximum level of: 182 g hydrogen peroxide and ultraviolet lamp power of 21 W. According to the plan performed, the level of titanium dioxide added to the process can be adjusted at low or medium level, i.e. with values 0.083g / L and 0.167g / L. According to Malik and Saha (2003), the influence of peroxide with temperature is related to the efficiency of ratio by using this compound and its rapid decomposition in the reaction.

Table 6. Results of replica 2 of the percentage reduction percentage obtained in experiments, initial TOC of 7920mg / l.

Experiment	pH Factor A	TiO2 Factor B	H2O2 Factor C	UV Factor D	Replica 2: reduction of total organic carbon (%)
1	1	1	1	1	31,269
2	1	2	2	2	34,498
3	1	3	3	3	36,443
4	2	1	2	3	35,720
5	2	2	3	1	31,648
6	2	3	1	2	30,019
7	3	1	3	2	30,739
8	3	2	1	3	33,011
9	3	3	2	1	27,481

Statistical analysis at a level of 95%, showed the most significant factors for the removal of organic load. According to the distribution F whose critical value is 4.26 and a p-value less than 5%, the most important factors for the degradation of organic matter in the effluent were phenolic H_2O_2, pH, UV. The most significant factor was the ultraviolet lamp power with F of 60.65201 and a p-value less than 0.001%, and then the remaining factors are the pH (F = 30.11586; p-value = 0.10 %) and H_2O_2 (F = 4.67497; p-value = 4.053%). The values obtained by analysis of variance confirmed the significance shown in the graph of main effects.

The analysis of variance (ANOVA) with F> 2 demonstrated that the factor Ti02 is significant for TOC removal. According to Phadke (1989), an F value statistically greater than 2 is considered as a significant effect (factor). The statistical significance factor in TOC reduction in the effluent treatment was confirmed by ANOVA, as shown in Table 7.

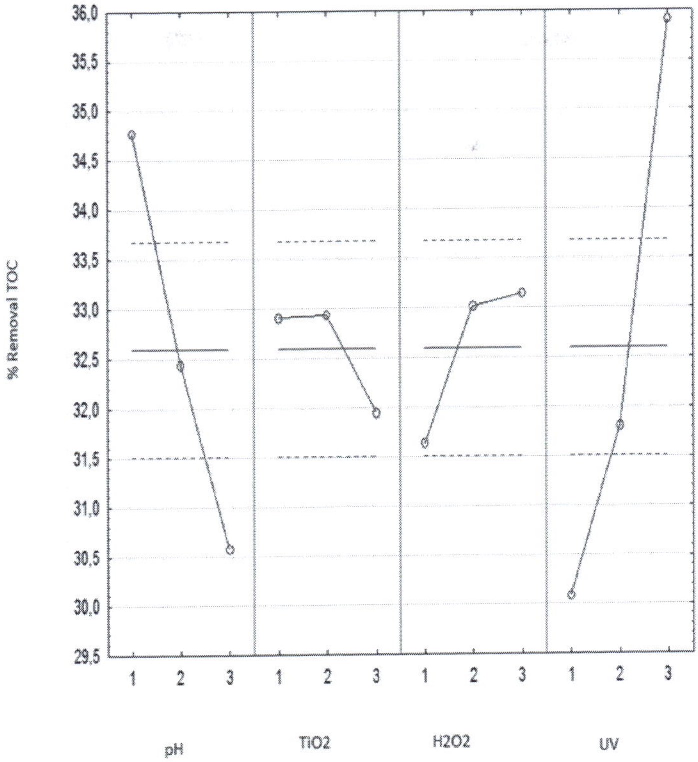

Figure 6. Main Effects in TOC percentage variation measurements in the effluent treatment of L_9 planning

Table 7. Analysis of variance Taguchi L_{16} orthogonal array obtained for TOC (%) removal

Source of Variation	SQ	GL	SMQ	F	P-Value
pH	53,0700	2,00	26,53499	30,11586	0,00010
TiO2	3,7711	2,00	1,88554	2,13999	0,17366
H2O2	8,2382	2,00	4,11909	4,67497	0,04053
UV	106,8806	2,00	53,44030	60,65201	<0,00001
Error	7,9299	9,00	0,88110		

Multiple linear regressions provide another statistical approach to evaluate variables in a quantitative approach. Significant parameters to the regression analysis are shown in Table 8, where pH, UV and H_2O_2 are relevant for the degradation process of the organic load of the polyester resin effluent.

Table 8. Regression parameters

Factor	Coefficient	t-value	Coef. Beta	Probability
pH	-1,04933	-4,573584	-0,542040	0,0003
TIO2	-5,73042	-1,042771	-0,123584	0,1580
H2O2	0,0266726	1,627560	0,192891	0,0638
UV	0,245688	5,791488	0,686380	0,0000
Constant	32,2247			

Table 9. ANOVA of multiple linear regression

Sources of Variation	GL	SQ	SMQ	F	P-Value
Due to Regression	4	147,0425	36,76063	14,55	0,0001
Independent	13	32,84718	2,526706		

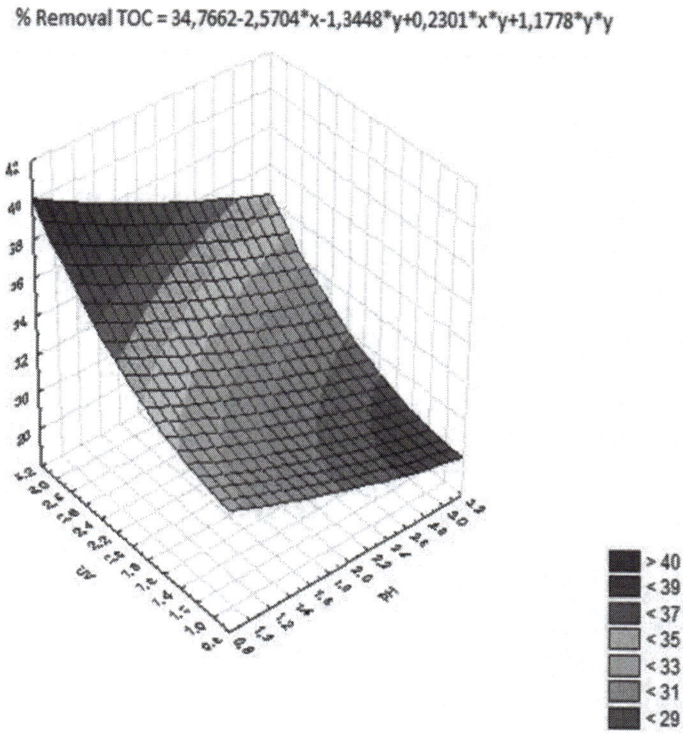

Figure 7. Graph of the two most influential factors in the process

Multiple linear regression showed a coefficient of determination (R2) of 0.817404, which demonstrates the efficiency of the degradation of effluent using the polyester resin of the experimental design. An ANOVA (Table 9) was performed in order to validate the multiple linear regression equation. The significance of equation shown is for a level of significance equal to 0.0001 at 95% confidence degree.

The most influential factors in the process show the percentage removal of Total Organic Carbon inFigure 7. An increasing degradation of the organic load is observed on the surface. This is achieved by independent variables: pH and potency of the ultraviolet lamp. The percentage of the increased removal of organic load occurs when there is an increase of the power of the lamp and a decrease of the pH. The greatest percentage reduction of the organic load is equal to 39.489%, whose parameters used in this experimental condition = H2O2 were 182 g, pH = 3, TiO2 = 0.250 g / L and the lamp power of 21 W. The response variable was significant for the degradation of organic matter in the effluent.

11. CONCLUSIONS

Taguchi planning was applied to the degradation of effluent organic load. The experimental design showed that further reduction of TOC (%) is related to an increase in pH and ultraviolet intensity. The results obtained were significant for the removal of TOC (%) from polyester resin effluent treated with advanced oxidation processes, and heterogeneous photocatalysis.

In our Taguchi orthogonal array the removal-percentage achieved was COT = 39.489%, which corresponds to experimental condition number three. This condition is inclusive of the weight ratio of hydrogen peroxide at 183g, pH = 3, TiO 2 = 0.250 g/L and the lamp intensity = 21 W. We conclude that the process of heterogeneous photocatalysis is optimally suitable for treatment of the effluent studied in this work.

REFERENCES

1. AbrafatiPaints & Coatings: Science and Technology. 2nd ed. Sao Paulo: New text- Publisher And Editorial Services Ltd, 1995

2. F Ay, E C Catalkaya, F Kargi, A Statistical Experiment Design Approach for Advanced Oxidation of Direct Red Azo Dye by Photo-Fenton Treatment. Journal of Hazardous Materials 2009;1621230236http://www.ncbi.nlm.nih.gov/pubmed/18555599accessed 03 July 2012
3. N Barros, R E Bruns, I S Scarminio, Optimization and Planning of Experiments. Campinas: Publisher of Unicamp;1995
4. K R Beebe, R J Pell, M B Seasholtz, Chemometrics: A Pratical Guide. New York: John Wiley and Sons; 1998
5. R E Bruns, B B Neto, I S Scarminio, How Experiments. Porto Alegre: Publisher Artmed; 2010
6. G Busca, S Berardinelli, C Resini, L Arrighi, Technologies for the removal of phenol from fluid streams: A short review of recent developments. Journal of Hazardous Materials 2008;160 (2-3) 265-288. http://www.sciencedirect.com/science/article/pii/S0304389408004172accessed 03 July 2012
7. Carp, C L Huisnab, A Reller, Photoinduced reactivity of titanium dioxide-progress in solid state chemistry. Progress in Solid State chemistry 2004; 32(1-2) 33-177. http://www.sciencedirect.com/science/article/pii/S0079678604000123accessed July 2012
8. R Ciola, Foundations of Catalysis. São Paulo: University of São Paulo; 1981
9. G Custo, M I Litter, D Rodríguez, C Vàsquez, Total reflection X-ray fluorescence trace Mercury determination by trapping Complexation: Application Technologies in Adavanced oxidation. Spectrochimica Acta Part B 2006;1119-1123.
10. X Domenèch, W F Garden, M. I Litter, Avanzados Oxidation Processes for La eliminación contaminants. In: Contaminant Eliminacion. La Plata: Network CYTED, 2001
11. G Franceschini, S Macchietto, Model-based design of experiments for parameter precision: State of the art. Chemical Engineering Science 2008, 68(19) 4846-4872. http://www.sciencedirect.com/science/article/pii/S0009250907008871accessed 03 July 2012
12. S H Gabardo, Study and design of photochemical reactors for wastewater treatment chemicals. Dissertation (Masters in Chemical Engineering) School of Chemical Engineering, State University of Campinas, Campinas; 2005

13. F Han, R Kambala, S V Srinivasan, M Rajarathan, D Naidu, R. Tailored titanium dioxide photocatalysts for the degradation of organic dyes in wastewater treatment: A review. Applied Catalysis A: General 2009, 359 (1-2) 25-40.http://www.sciencedirect.com/science/article/pii/S0926860X09001756accssed 04 July 2012
14. P K Hopke, The evolution of ChemometricsAnalytical Chemica Acta, 3653772003Review).
15. H Kusic, N Koprivanac, L Srsan, Azo dye degradation using Fenton type processes assisted by UV irradiation: A kinetic studyJournal of Photochemistry and photobiology A: Chemistry 2007195 EOF202 EOF
16. A Lermontov, L Yokoyama, M Lermontov, B Machado, U T. Application of fuzzy logic in the parameterization of a new index of water quality. Directory of Open Access Journals 2008
17. S H Lin, R S Juang, Adsorption of phenol and its derivates from water using synthetic resins and low-cost natural adsorbents: A review.Journal of Environmental Management 2009, 90(3) 1336-1349.
18. P K Malik, S K Saha, Oxidation of direct dyes with hydrogen peroxide using ferrous ion the catalyst. Separation and Purification Technology 2003, 31(3) 241-250.
19. L Metcalf, H P Eddy, Wastewater Engineering: Treatment, Disposal and Reuse. São Paulo: McGraw-Hill, 1991
20. D C Montgomery, Design and Analysis of Experimentsth ed. New York: John Wiley and Sons, 2001
21. J L Morais, Study of the potential of advanced oxidation processes, isolated and integrated with traditional biological processes, for tratmento landfill leachate. 229 f. Thesis (Ph.D. in Chemical Engineering)- Department of Chemistry, Federal University of Paraná, Curitiba, 2005
22. M Nehdi, J Sumner, Recycling waste latex paint in concreteCement and Concrete Research 2003, 33(6) 857-863.
23. R Nogueira, F P Jardim, W F. The heterogeneous photocatalysis and environmental applications. New Chemistry 1998
24. J H Oliver, K Hynook, C Pen-chien, Decolorization of wastewater, Crit. Rev. Environmental Science Technology 2000
25. I Oller, S Malate, J A Sanchez-perez, Combination of the advanced oxidation process and treatments for biological waste water decontamination: A review. Science of the total environment 2011

26. E Ongley, Modernization of water quality programs in Developing coutries: issues of relevancy and costefficiency. Water Quality International Sep / Oct 1998, p. 37-42.
27. M Otto, Modeling. In: Chemometrics: Statistics and computer application in analytical chemistry.New York: Wiley, 2007
28. A P Paiva, Response surface methodology and principal component analysis for optimization of manufacturing processes with multiple correlated responses. 129 f. Thesis (PhD in Mechanical Engineering)- Institute of Mechanical Engineering, Federal University of Itajubá, Itajubá, 2006
29. A P Paiva, J R Ferreira, P P Balestrassi, A hybrid multivariate approach applied to AISI 52100 hardened steel turning optimizationJournal of Materials Processing Technology200726 EOF
30. M. S Phadke, Quality engineering using robust designPublisher Prentice Hall. 1989
31. T Rajalahti, O. M Kvalheim, Multivariate data analysis in pharmaceutics: A tutorial reviewInternational Journal of Pharmaceutics2011280 EOF290 EOF
32. J L Rosa, A Robin, M B Silva, C A Baldan, M P Peres, Electrodeposition of copper on titanium wires: Taguchi Experimental Design ApproachJournal of Materials Processing Technology 2009, 209 (3) p. 1181-1188.
33. R Salazar, F S. Application of advanced oxidation processes and pre-treatment ofdairy effluent for subsequent biological treatment. 210 f. Dissertation (Masters in Chemical Engineering)- Engineering School of Lorena, University of São Paulo, Lorena 2009
34. Y Segura, R Molina, F Martínez, J A Melero, Integrated sleep heterogeneous photo-Fenton processes for the degradation of phenolic aqueous solutions. Ultrasonics Sonochemistry 2009, 16 417-424.
35. P Sharma, A Verma, R K Sidhu, O P Pandey, Process parameter selection for strontium ferrite sintered magnets using Taguchi L9 orthogonal designJournal of Materials Processing Technology2005, 168 (1) p.147-151.
36. M Steiner, T A Neto, A C Braulio, S N Alves, V. Multivariate statistical methods applied to engineering evaluations. Management & Production, San Carlos 2008

37. R Thiruvenkatachari, T O Kwon, J C Jun, S Balaji, M Matheswaran, I. S Moon, Aplication of Several advanced oxidation process for the destruction of terephthalic acid (TPA). Journal of Hazardous Materials, v.142, p.308-314, 2007.

38. T Y Wang, C Y Huang, Improving forecasting performance by Employing the Taguchi methodEuropean Journal of Operational Research20071052 EOF1065 EOF

39. K D Weiss, Paint and Coatings: a mature industry in transitionProg. Polym. Sci, v. 22, p. 203-245, 1997.

5

CHAPTER

Application of Multivariate Data Analyses in Waste Management

K. Böhm[1], E. Smidt[2] and J. Tintner[2]

[1] Institute of Waste Management, Department of Water, Atmosphere and Environment, University of Natural Resources and Life Sciences, Vienna, Austria

[2] Institute of Wood Science and Technology, Department of Material Sciences and Process Engineering, University of Natural Resources and Life Sciences, Vienna, Austria

1. INTRODUCTION

First of all, what is multivariate data analysis and why is it useful in waste management?

Methods dealing with only one variable are called univariate methods. Methods dealing with more than one variable at once are called multivariate methods. Using univariate methods natural systems cannot be described satisfactorily. Nature is multivariate. That means that any particular phenomenon studied in detail usually depends on several factors. For example, the weather depends on the variables: wind, air pressure, temperature, dew point and seasonal variations. If these factors are collected every day a multivariate data matrix is generated. For interpretation of such data sets multivariate data analysis is useful. Multivariate data analysis can be used to process information in a meaningful fashion. These methods can afford hidden data structures. On the one hand the elements of measurements often do not contribute to the relevant property and on the other hand hidden phenomena are unwittingly recorded. Multivariate data analysis allows us to handle huge data sets in order to discover such hidden data structures which contributes to a better understanding and easier interpretation. There are many multivariate data analysis techniques available. It depends on the question to be answered which method to choose.

Due to the requirement of representative sampling number of samples and analyses in waste management lead to huge data sets to obtain reliable results. In many cases extensive data sets are generated by the analytical method itself. Spectroscopic or chromatographic methods for instance provide more than 1000 data points for one sample. Evaluation tools can be developed to support interpretation of such analytical methods for practical applications. For specific questions and problems different evaluation tools are necessary. Calculation and interpretation are carried out by the provided evaluation tool.

In this study an overview of multivariate data analysis methods and their application in waste management research and practice is given.

2. MULTIVARIATE DATA ANALYSIS IN WASTE MANAGEMENT

The main objectives of multivariate data analysis are exploratory data analysis, classification and parameter prediction. Many different multivariate data analysis methods exist in literature. Thus the following list is not exhaustive however subdivided into the mentioned superior categories. It only concentrates on the methods applied in waste management.

Table 1 gives an overview of the existing literature in waste management on multivariate data analysis applied by several authors. It can be summarised that PCA and PLS1 are the most popular multivariate data analysis methods applied in waste management. Details are given in the following sections 2.1 and 2.2. Due to easy traceability of the parameters investigated in the different papers parameter descriptions have been taken as they were mentioned in the original.

In practice there are many software packages available which include different multivariate data analysis methods. Some software tools are: SPSS (www.spss.com\de\statistics), Canoco (www.canoco.com), The Unscrambler (www.camo.com) and the Free Software R-project (www.cran.r-project.org).

2.1. Pattern Recognition

2.1.1. Exploratory data analysis

- Principal Componant Analysis (PCA)

PCA is mathematically defined as an orthogonal linear transformation that arranges the data to a new coordinate system in that the greatest variance by any projection of the data takes place along the first coordinate (called the first principal component), the second greatest variance along the second coordinate, and so on. Theoretically the PCA is the optimum transformation for a given data set in least square terms. That means PCA is used for dimensionality reduction of variables in a data set by retaining those characteristics of the data set that contribute most to its variance. The transformation to the new coordinate system is described by scores (T), loadings (P) and errors (E). In matrix terms, this can be written as X = T * P + E. Fig. 1 illustrates the mathematical transformation using PCA. The matrices can be displayed graphically. The scores matrix illustrates the data structure and the loading matrix displays the influence of the different variables on the data structure.

TABLE 1. Literature review of different multivariate data analysis methods applied in waste management; PCA – Principal Component Analysis, FA – Factor Analysis, CA – Cluster Analysis, CCA – Canonical Correspondence Analysis, DA – Discriminant Analysis, SIMCA – Soft Independent Modelling of Class Analogy, MLR – Multiple Linear Regression, PLS-R – Partial Least Squares Regression, PSR – Penalised Signal Regression

	Pattern recognition						Calibration			
Method	PCA	FA	CCA	CA	DA	SIMCA	MLR	PLS1	PLS2	PSR
Chapter		2.1.1		2.1.2	2.1.3	2.1.3	2.2.1		2.2.2	
Compost science	[1-23]	[24]	[25]	[1, 4, 22, 24-31]	[3, 9]	[8, 12]	[29, 32, 33]	[2, 6, 8, 19, 21, 23, 34-47]	[8, 21, 48]	[49]
Municipal solid waste	[50-55]			[56]				[17, 53, 57, 58]		
Landfill research	[59-72]	[65]	[73, 74]	[72, 75]	[66, 71, 76, 77]	[78]	[79, 80]	[17, 61, 62, 66, 71, 78]		
Logistics	[81]	[82]		[82]			[83, 84]			

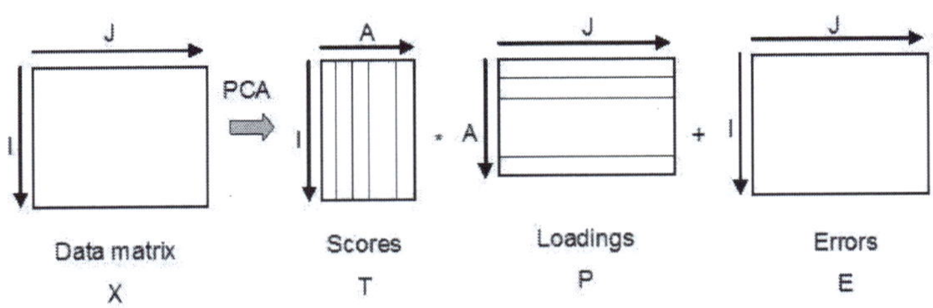

FIGURE 1. Principle of the PCA (according to Esbensen [85])

PCA displays hidden structures of huge data sets. PCA is applied in different fields of waste management to find out the relevant parameters of a large parameter set. So we can see which properties of a sample are significant and important to answer a particular question. Due to the results obtained time and money can be saved in further research activities.

Many applications can be found in compost science. Zbytniewski and Buszewski [1] applied PCA to reveal the significant parameters and possible groupings of chemical parameters, absorption band ratios and NMR data. Campitelli and Ceppi [3] investigated the quality of different composts and vermicomposts. The collected data were evaluated by means of PCA to extract the significant differences between the two compost types. Gil et al. [4] used PCA to show effects of cattle manure compost applied on different soils. Termorshuizen et al. [13] carried out a PCA based on disease suppression data determined by bioassays in different compost/peat mixtures and pure composts. PCA was applied by Planquart et al. [10] to examine the interactions between nutrients and trace metals in colza (Brassica napus) when sewage sludge compost was applied to soils. LaMontagne et al. [7] applied PCA on terminal restriction fragment length polymorphisms (TRFLP) patterns of different composts to reveal their characteristics with respect to microbial communities. Malley et al. [8] recorded near infrared spectra from cattle manure during composting. The collected spectral data were evaluated by PCA to show the relationships among samples and changes due to stockpiling and composting. Hansson et al. [6] observed the anaerobic treatment of municipal solid waste by using on-line near infrared spectroscopy. For spectral data interpretation PCA was carried out. Albrecht et al. [2] also performed a PCA for near infrared (NIR) spectra evaluation from an ongoing composting process. Smidt et al. [12] used PCA to show differences in spectral characteristics of different waste materials. Lillhonga et al. [23] used PCA to observe spectral characteristics of different composting processes. Vergnoux et al. [21] applied a PCA on NIR spectra as well as on physico-chemical and biochemical parameters to derive regularities from the data. Nicolas et al. [9] used PCA to evaluate data from an electronic nose. The correlations between the sensor of an electronic nose and chemical substances were determined by Romain et al. [11] using PCA. PCA was applied to observations of a composting process by means of analytical electrofocusing. The electrofocusing profiles were evaluated by Grigatti et al. [5]. PCA was also used by Biasioli et al. [19] to evaluate odour emissions and biofilter efficiency in composting plants using proton transfer reaction-mass spectrometry. Bianchi et al. [18] also used PCA to reduce the complex data set and to analyse the pattern of organic compounds emitted from a composting plant, a municipal solid waste landfill and ambient air. The effect of 14 different soil amendments on compost quality were evaluated using a PCA by Tognetti et al. [20]. Smidt et al. [16] applied PCA to illustrate the influence of input materials and composting operation on humification of organic matter. Böhm et al. [14] and Smidt et al. [15, 17] used PCA to illustrate spectral differences caused by different materials such as biowaste, manure, leftovers, straw and sewage sludge.

PCA was also applied to illustrate the alteration of municipal solid waste during the biological degradation process reaching stability limits for landfilling as well as to demonstrate similarities and differences of reactor and old landfills based on thermal data [53, 66]. Scaglia and Adani [52] focused on municipal solid waste treatment. They used PCA to create a stability index for quantifying the aerobic reactivity of municipal solid waste. Abouelwafa et al. [54, 55] investigated the degradation of sludge from the effluent of a vegetable oil processing plant mixed with household waste from landfill. Abouelwafa et al. [54] applied PCA on various parameters measured during composting (e.g. pH, electrical conductivity, moisture, C/N, NH_4/NO_3, ash, decomposition in percent, level of polyphenols, lignin, cellulose, hemicellulose, humic acid) to find the main parameters in the decomposition and restructuring phase [54]. Abouelwafa et al. [55] extracted fulvic acids from the samples mentioned above and extended the data set used for PCA by a series of absorption band ratios resulting from of FTIR spectra.

PCA has also been used in landfill research. Mikhailov et al. [62] applied PCA for monitoring data from different landfills. They included parameters such as depth, ash content, volumetric weight, humidity, amounts of refuse in summer and winter as well as the topsoil depth of landfill sections, sewage sludge lenses and the existence of a protection system. Kylefors [61] investigated data of leachate composition using PCA. The idea was to reduce the analytical monitoring program for further investigations. Durmusoglu and Yilmaz [60] used PCA to extract the significant independent variables of the collected data of raw and pre-treated leachate. A comparable work was done by De Rosa et al. [59]. They also investigated the leachate composition of an old waste dump connected to the groundwater. Olivero-Verbel et al. [63] investigated the relationships between physico-chemical parameters and the toxicity of leachates from a municipal solid waste landfill. PCA was used to find out which parameters were responsible for their toxicity. Jean and Fruget [72] used PCA to compare landfill leachates according to their toxicity and physico-chemical parameters. Ecke et al. [71] showed an example for PCA application in landfill monitoring of data from landfill test cells, leachate and gas data. Smidt et al. [64] investigated landfill materials by means of mid infrared spectroscopy, thermal analysis and PCA. They used PCA to support data interpretation. Van Praagh et al. [70] investigated the potential impacts on leachate emissions using pretreated and untreated refuse-derived material as a cover layer on the top of a municipal solid waste landfill. To interpret leachate characteristics they used PCA. Tintner and Klug [69] used PCA to illustrate how vegetation can indicate landfill cover features. Diener et al. [67] investigated the long-term stability of steel slags used as cover construction of a municipal solid waste landfill by means of a PCA. Smidt et al. [17] used PCA to display spectral characteristics of different landfill types.

Pablos et al. [68] used a PCA to evaluate toxicity bioassays for biological characterisation of hazardous wastes.

Other publications focus on the process monitoring of municipal solid waste incineration residues. Ecke [50] performed PCA on leaching parameters from municipal solid waste incineration fly ash to get an overview of the mobility of metals under certain conditions. Mostbauer et al. [51] carried out PCA to observe the long-term behaviour of municipal solid waste incineration (MSWI) residues.

In the field of waste management logistics PCA is rarely applied. Dahlén et al [81] used PCA to display the impact of waste costs on a weight basis in a specific municipality.

- Factor Analysis (FA)

FA is related to PCA but differs in its mathematical conception [86]. FA is also used to describe the variability of observed variables in terms of fewer variables called factors. That means factor analysis is a tool which reveals unobservable underlying features of a specific phenomenon by previous visible observations. The observed variables are modelled as linear combinations of the factors plus "error" terms. The information about interdependencies can be used to reduce the number of variables in a data set.

In waste management practice PCA is preferentially used. Differences between factor analysis and PCA are found to be small [86]. Srivastava and Ramanathan [65] investigated the groundwater quality of a landfill site in India by means of FA. They explained the observed relationship in simple terms expressed as factors. Bustamante et al. [24] used FA to identify the principal variables associated to the composting of agro-industrial wastes. Lin et al. [82] used FA for selecting the best food waste recycling method.

- Canonical Correspondence Analysis (CCA)

CCA is a multivariate method to explain the relationships between biological communities and their environment [87]. The method is designed to extract environmental gradients from ecological data sets. By means of the gradients an ordination diagram describing and visualising the diverse habitat preferences of taxa is calculated.

CCA is sometimes used in waste management if, for example, microbial communities or vegetation surveys are analysed. CCA was applied by Franke-Whittle et al. [25] and El-Sheikh et al. [73]. Franke-Whittle et al. [25] applied CCA to illustrate the similarities in microbial communities of three different composting processes. El-Sheikh et al. [73] investigated the ten-year primary succession on a newly created landfill at a lagoon of the Mediterranean Sea. Vegetation surveys where the basis for CCA. Kim et al.

[74] applied CCA to investigate the vegetation and the soil of a not properly maintained landfill to suggest restoration alternatives by comparing the vegetation of the landfill to the nearby forests.

2.1.2. Unsupervised pattern recognition

- Cluster analysis (CA)

Clustering is the classification of objects into groups called clusters. Objects from the same cluster are more similar to one another than objects from different clusters. The difference of clusters is based on measured distances without any unit. Cluster analysis can be illustrated graphically in a dendrogram as shown in Fig. 2. The samples 2, 3 and 5 are clustered due to the high degree of similarity as well as the samples 1 and 4. The two clusters show little similarity.

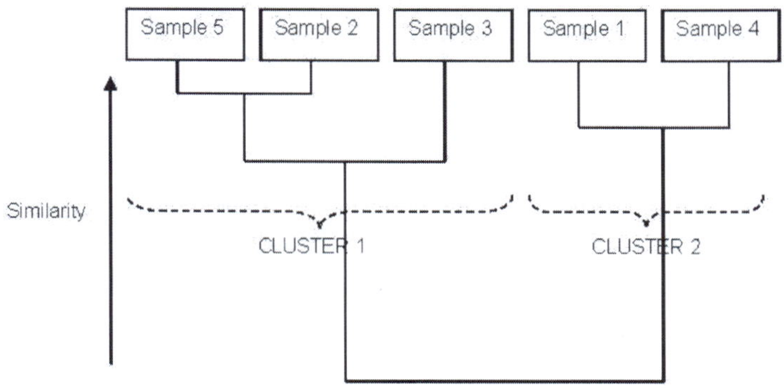

FIGURE 2. Example of a cluster analysis visualised by a dendrogram

CA was applied in compost science by Zybtniewskie and Buszewski [1]. They applied CA to conventional compost parameters and NMR data to find out the grouping depending on the composting time. He et al. [56] used a hierarchical cluster analysis to show the similarities and differences of UV-Vis and fluorescence spectra of water extractable organic matter, originating from municipal solid waste that had been subjected to different composting times. A hierarchical cluster analysis was also used by He et al. [22] to investigate water-extractable organic matter during cattle manure composting. Gil et al. [4] displayed dendrograms to illustrate the similarities or differences by application of cattle manure compost to different soils. Bustamante et al. [24] studied physico-chemical, chemical and microbiological parameters of different composts. The evaluation of the composts was conducted by a hierarchical cluster analysis [24].

Lin et al. [82] applied a CA for the selection of optimal recycling methods for food waste.

A stepwise cluster analysis (SCA) was used to describe the nonlinear relationships among state variables and microbial activities of composts by Sun et al. [29]. Sun et al. [30] developed a genetic algorithm aided stepwise cluster analysis (GASCA) to describe the relationships between selected state variables and the C/N ratio in food waste composting.

Furthermore CA has often been used to evaluate microbiological data, especially in compost science [25-28, 31]. Innerebner et al. [26] and Ros et al. [27, 28] used CA to identify related samples and similar groups of microorganisms. Franke-Whittle et al. [25] used CA to show the similarities of Denaturing Gradient Gel Electrophoresis (DGGE) data of three different compost types with proceeding compost maturity. Xiao et al. [31] used a hierarchical cluster analysis of DGGE data to estimate the succession of bacterial communities during the active composting process.

Tesar et al. [75] applied CA to spectral data to illustrate the effect of in-situ aeration of a landfill. Jean and Fruget [72] used CA to compare landfill leachates on the basis of their toxicity and physico-chemical parameters.

2.1.3. Supervised pattern recognition

All supervised methods are classifications. Classification can be considered as a predictive method where the response is a category variable. Different classification methods exist. There are types of "hard" and "soft" modelling. Hard modelling means that a non-relocatable line between the defined groups exists. One object can only belong to one group. Soft modelling allows an overlapping of the defined classes. An object can belong to both groups [88]. With regard to waste management practice two different classification methods are described in detail.

- Discriminant analysis (DA)

DA is a classification method of hard modelling. Campitelli and Ceppi [3] carried out a DA to distinguish between compost and vermicompost on the basis of parameters such as total organic carbon (TOC), germination index (GI), pH, total nitrogen (TN), and water soluble carbon (WSC). Nicolas et al. [9] performed a DA to classify data of an electric nose according to defined exceeded levels of odour. Ecke et al. [71] investigated samples from three different landfill sites by the biochemical methane potential and used DA for data evaluation. Huber-Humer et al. [77] applied DA to determine methane oxidation efficiency of different materials based on chemical and physical variables. Smidt et al. [66,76] used DA to differentiate the infrared spectral [76] and thermal patterns [66] of municipal solid waste incinerator (MSWI) bottom ash before and after CO_2 uptake. A DA on the CO_2 ion current recorded during combustion was applied to illustrate the effect of CO_2 treatment of MSWI bottom ash [66]. DA was also used to illustrate the spectral characteristics of leachate from landfill simulation reactors under aerobic and anaerobic conditions [17].

- Soft independent modelling of class analogy (SIMCA)

SIMCA is a special method of soft modelling recommended by Wold in the 1970s [88]. Objects can belong to one of the defined class, to both classes or to none. Whether SIMCA can be applied on the data set depends on the question to be answered. According to Brereton [88] it is often legitimate in chemistry that an object belongs to more than one class For example a compound may have an ester and an alkene group which are both reflected by an infrared spectrum. Thus they fit in both classes. In natural science it is allowed in most cases for an object to be in line with more than one class simultaneously.

Contrarily in other cases an object can belong only to one class and the application of SIMCA is inappropriate. Brereton [88] gives a good example where the concept of SIMCA is not applicable: A banknote is either forged or not. In many cases there is only one true answer. For such problems SIMCA is not the adequate method.

In compost science Malley et al. [8] and Smidt et al. [12] carried out a SIMCA. Malley et al. [8] classified different decomposition stages of manures by means of near infrared spectroscopy and SIMCA. Smidt et al. [12] carried out a SIMCA to classify different waste materials such as biowaste compost, mechanically-biologically pretreated waste and landfill materials based on their spectroscopic pattern. Smidt et al. [78] used the SIMCA model developed by Smidt et al. [12] to identify different landfill types such as reactor landfill and industrial landfill samples.

2.2. Calibration

2.2.1. Multiple linear regression (mlr)

MLR is directed at modelling the relationship between two or more explanatory variables and a response variable by fitting a linear equation to observed data. Every value of the independent variable X is associated with a value of the dependent variable Y, with explanatory or predictive purposes. A direct correlation between Y and X-matrix is performed.

In waste management MLR was applied by Chikae et al. [32] to predict the germination index which was adopted as a marker for compost maturity. Thirty-two parameters of 159 samples were measured. MLR was carried out to reduce this huge parameter set to some significant parameters. Lawrence and Boutwell [79] used MLR for predicting the stratigraphy of landfill sites using an electromagnetic method. Moreno-Santini et al. [80] applied MLR to determine arsenic and lead levels in the hair of residents in a municipality constructed on a former landfill.

Noori et al. [84] compared two different statistical methods (artificial neural networks and MLR based on a PCA) to predict the solid waste generation in Tehran. Cheng et al. [83] used MLR to predict the factors associated with medical waste generation at hospitals. Sun et al. [29] used MLR to predict mesophilic and thermopilic bacteria in food waste composts. Suehara and Yano [33] applied MLR to predict conventional compost parameters by NIR spectral data.

2.2.2. Partial least squares regression (pls-r)

- PLS1

PLS-R is used to find out the fundamental relations between two matrices. PLS-R is a bilinear modelling method. The main idea behind it is to calculate the principal components of the X and the Y matrix separately (external correlation) and to develop a regression model between the scores of the principal components (inner correlation). The concept of PLS-R is demonstrated in Fig. 3.

PLS1 is often used to predict time consuming or expensive parameters using an alternative analytical method. Modern analytical tools such as spectroscopic, chromatographic and thermo analytical methods generate data with inherent information on different parameters. With the development of an evaluated prediction model conventional analytical methods can be replaced by easier and/ or faster handling and robust methods.

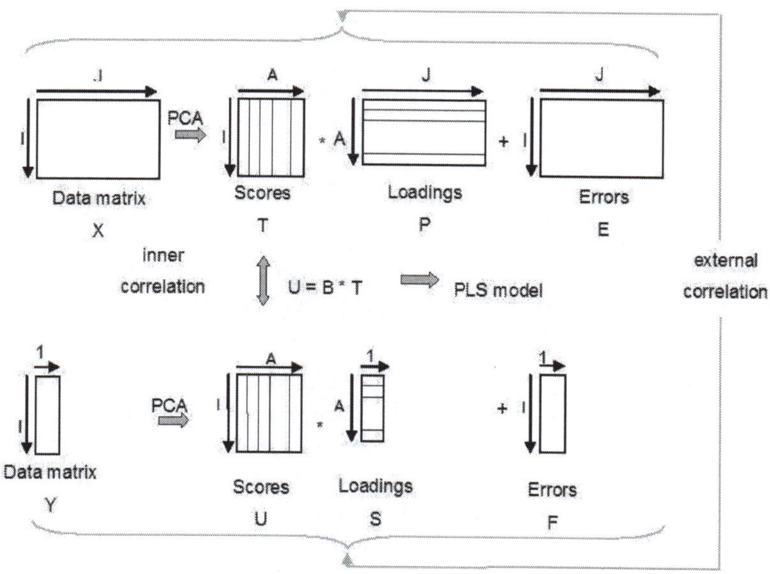

FIGURE 3. Principles of PLS-R (according to Esbensen [85])

Many authors have developed such prediction models in compost science. Zvomuya et al. [44] predicted phosphorus availability in soils, amended with composted and non-composted cattle manure by means of cumulative phosphorus analysis. Fujiwara and Murakami [35] applied near infrared spectroscopy to estimate available nitrogen in poultry manure compost. Huang et al. [36] also used near infrared spectroscopy to estimate pH, electric conductivity, volatile solids, TOC, total N, the C:N ratio and the total phosphorus content. Furthermore they determined nutrient contents such as K, Ca, Mg, Fe and Zn of animal manure compost using near infrared spectroscopy and PLS1 [37]. Malley et al. [8] developed prediction models for total C, organic C, total N, C:N ratio, K, S and P by means of near infrared spectroscopy and PLS1. Morimoto et al. [43] carried out carbon quantification of green grass tissue using near infrared spectroscopy. Hansson et al. [6] predicted the concentration of propionate in an anaerobic process by near infrared spectra. Albrecht et al. [2] developed calibration models between spectral data and C, N, C:N ratio and composting time. Michel et al. [42] predicted chemical and biological properties of composts such as organic C (C_{org}), total N, C:N ratio, age, microbial biomass (C_{mic}), C_{mic}:C_{org}, basal respiration, enzymatic activity and plant suppression using near infrared spectroscopy. Ludwig et al. [39] also used near infrared spectroscopy to predict pH, electric conductivity, P, K, NO_3^- and NH_4^+ and phytotoxicity. Ko et al. [38] predicted heavy metal contents of Cr, As, Cd, Cu, Zn and Pb by means of near infrared spectroscopy and PLS1. They hypothesised that heavy metals are detectable by NIR when they are complexed with organic matter. Capriel et al. [34] found out that mid infrared spectroscopy is a rapid method to estimate the effect of nitrogen and relevant parameters such as total C, total N, the C:N ratio and the pH of biowaste compost. Meissl et al. [40] used PLS1 and the mid infrared region to predict humic acid contents in biowaste composts. Furthermore they determined humic acid contents by near infrared spectroscopy [41]. Sharma et al. [47] developed prediction models for conventional compost parameters, especially ammonia, pH, conductivity, dry matter, nitrogen and ash using NIR and Vis-NIR spectroscopy. Lillhonga et al. [23] used PLS-R for compost parameter prediction based on NIR spectra. They developed models for the parameters: time, pH, temperature, NH_3/NH_4^+, energy (calorific value) and moisture content. Galvez-Sola et al. [45] used PLS1 to predict different compost quality parameters such as pH, electric conductivity, total organic matter, total organic carbon, total N, C/N ratio as well as nutrients contents (N, P, K) and potentially pollutant element concentrations (Fe, Cu, Mn and Zn) from near infrared spectra. Vergnoux et al. [21] applied a PLS1 to predict physico-chemical and biochemical parameters from NIR spectra. Physico-chemical parameters comprised age, organic carbon, organic nitrogen, C/N, total N, fulvic acids (FA), humic acids (HA) and HA/FA. The soluble fraction, lignin and biological maturity index were summarised as biochemical parameters. Mikhailov et al. [62] used PLS1 to predict maturity and stability based on conventionally measured data. Kylefors [61] developed prediction models for leachate concentrations of specific organic substances in leachate by means of conventional leachate analysis and PLS1. Biasioli et al. [19] used PLS1 to predict odour concentrations in composting plants by proton transfer reaction-mass spectrometry (PTR-MS). Mohajer et al. [46] used a PLS1 to generate a model to predict the microbial oxygen uptake in sludge based on different physical compost parameters.

Böhm et al. [57] used PLS1 to predict the respiration activity (RA_4) based on FT-IR spectra of mechanically-biologically pretreated (MBT) waste. The potential of thermal data of MBT waste was shown by Smidt et al. [53]. They applied PLS1 to predict the calorific value, total organic carbon (TOC) and respiration activity (RA_4). Smidt et al. [17] also developed a prediction model for the calorific value based on spectral data. Biasioli et al. [58] used PLS1 to predict odour concentration from MSW composting plants based on PTR-MS.

Ecke et al. [71] performed detoxification of hexavalent chromium to less toxic trivalent chromium in industrial waste and applied a PLS model to identify the relevant factors. Smidt et al. [78] predicted the biological oxygen demand and the dissolved organic carbon (DOC) of old landfill materials from spectral data. They also used PLS-R to predict the total organic carbon and total nitrogen based on thermal data [78]. Furthermore PLS-R was used to predict respiration activity (RA_4) from MS data of old landfill materials [66]. Smidt et al. [17] developed a prediction model for the DOC and the TOC from spectral data of landfill materials.

- PLS2

PLS2 is a variant of the PLS-R method where several Y-variables are modelled simultaneously. An advantage of this method is to find possible correlations or co-linearity between the Y-variables.

Malley et al. [8] developed prediction models for pH, total N, nitrate and nitrite, total C, organic C, C:N ratio, P, available P, S, K and Na by means of near infrared spectroscopy and PLS2. Suehara et al. [48] used PLS2 for simultaneous measurement of carbon and nitrogen content of composts using near infrared spectroscopy. Vergnoux et al. [21] applied PLS2 to predict physico-chemical (moisture, temperature, pH, NH_4-N) and biochemical parameters (hemicellulose and cellulose) from NIR spectra.

- Penalised signal regression (PSR)

This special regression method is described in Galvez-Sola et al. [49]. Galves Sola et al. [49] used this method to predict the phosphorus content in composts.

3. SELECTED EXAMPLES FROM LITERATURE USING MULTIVARIATE DATA ANALYSIS IN WASTE MANAGEMENT

In the following chapter four selected examples using multivariate data analysis in waste management are described in detail. To illustrate the application of principal component analysis (PCA) the study by Mikhailov et al. [62] is presented. He carried out multivariate data analysis for the ecological assessment of landfills. The second example illustrates the application of partial least squares regression (PLS-R). Michel et al. [42] applied PLS-R to predict conventional parameters by spectroscopic data. Ros et al. [27] applied a cluster analysis to data of polymerase chain reaction coupled with denaturing gradient gel electrophoresis (PCR-DGGE) to observe the long-term effects of compost amendment on soil microbial activity. A soft independent model of class analogy (SIMCA) was applied by Malley et al. [8]. They used SIMCA to classify different composts according to their spectroscopic characteristic.

3.1. Principal Component Analysis (PCA)

3.1.1. Objective of the study

The objective of the study by Mikhailov et al. [62] was to evaluate the stability of landfills based on many conventional parameters such as ash content, temperature, volume weight, pH, humidity and depth. They supposed that a multivariate approach could provide a more efficient data interpretation. Therefore they compared conventional and multivariate data analysis methods.

3.1.2. Method of evaluation and results

In a first step Mikhailov et al. [62] collected conventional data to describe landfill stability. They investigated 3 different landfills in Russia, one illegal dump, an old poorly-run dump and a modern well-run landfill. They focused on geodesic surveys to obtain the overall object properties such as size, volume and different layers. Furthermore they investigated the physical and chemical properties of the samples collected in different depths of the landfill. The physical and chemical properties include ash content, humidity, and acidity. Using the conventional collected data they carried out a PCA for each landfill site. They included the ash content, temperature, volume weight, pH, humidity and depth. The PCA for the two landfills in Bezenchuk and Kinel are presented in the study [62]. Based on the data pool Mikhailov et al. [62] could identify two important sources of waste around Bezenchuk, a poultry farm and a granary. In addition to regular domestic refuse, the agricultural and industrial wastes were disposed illegally in this dump. Kinel on the other hand is a modern, well operated landfill, in which both domestic and industrial wastes are disposed. These assumptions were confirmed by chemometric investigations based on PCA. The mentioned PCAs show clustering of the different classes. The results of the PCA of the third investigated landfill are not shown in their study. Otradny was shown to be a poorly maintained landfill. Clear separation of layers by means of the scores plot was not possible. They found out that the information by the landfill manager and the results obtained did not correspond.

3.1.3. Conclusion

Mikhailov et al. [62] concluded that multivariate data analysis is an appropriate tool for ecological monitoring. They pointed out that chemometric methods provide the possibility to explore the structure of waste disposal by identification of specific areas.

3.2. Partial Least Square Regression (PLS1)

3.2.1. Objective of the study

The verification of compost quality has to be monitored consistently. However this is time-consuming and laborious. Due to the fact that NIR is a simple, accurate and fast technique used for routine analysis Michel et al. [42] hypothesised that NIR could be used for parameter prediction. The objective of the study was to use NIR spectroscopy to determine chemical and biological properties.

3.2.2. Method of evaluation and results

The first step was to define compost quality. Michel et al. [42] defined compost quality by C and N contents, suppression of pathogens, stability/ maturity and biological parameters, especially organic carbon (C_{org}), total N (N_t), C:N ratio, age, microbial biomass (C_{mic}), C_{mic}:C_{org}, basal respiration, enzymatic activity and suppression of plant disease. Spectroscopic data from 98 composts samples as well as the mentioned conventional parameters were collected. Fundamental relations between two matrices can be found by means of PLS1. Michel et al. [42] applied a PLS1 to express conventional parameters by spectral data. They designed for each conventional parameter a PLS1. Table 2summarises the collected data and results obtained by Michel et al. [42]. The standard error of cross-validation (SECV) and the coefficient of determination (r^2) indicate the quality of prediction. The SECV provides information on the prediction error, r^2 demonstrates the quality of correlation. Composting age and basal respiration show the highest r^2. The specific enzymatic activity and the suppressive effect show the lowest r^2. It should be emphasised that biological tests that are carried out with the original wet compost are more susceptible to interferences due to the heterogeneity of the material. Michel et al. [42] concluded that especially compost age and basal respiration are clearly reflected by the NIR spectrum and feature the best results. By contrast, the specific enzyme activity and suppressive effects show the worst prediction results. The assigned correlations are illustrated in the paper [42].

TABLE 2. Excerpt of table 1 and 2 by Michel et al. [42], SECV = standard error of cross-validation, r^2 = the coefficient of determination

	n	Mean	Range	Outliers removed	SECV	r^2
Age [d]	98	183.6	82.0 - 268.0	6	16.7	0.82
C_{org} content [%]	97	26.0	16.4 - 41.5	5	2.32	0.77
N_t content [%]	97	1.4	1.0 - 2.1	4	0.11	0.67
C:N ratio	97	18.2	12.2 - 29.1	4	1.51	0.71
C_{mic} [$\mu g\ g^{-1}$]	98	4986	774 - 8587	5	954	0.68
C_{mic}:C_{org} [mgCmicgCorg^{-1}]	97	18.6	4.0 - 29.4	4	4.00	0.63
Basal respiration [$\mu g\ C\ g^{-1}\ d^{-1}$]	47	574.8	252.0 - 966.0	2	49.2	0.88
qCO_2 [μgCO_2-C mg C_{mic}^{-1} d^{-1}]	47	9.7	4.2 - 17.1	1	1.98	0.83
Hydrolysis of fluorescein diacetate (FDA-HR) [$\mu g\ g^{-1} h^{-1}$]	98	517.9	256.0 - 879.0	5	74.7	0.75
Specific enzyme activity [μgFDA mg$C_{mic}^{-1} h^{-1}$]	98	118.7	48.6 - 370.9	6	48.6	0.49
Suppression 5‰ (rating) [%]	98	57.3	8.0 - 101.0	2	19.3	0.71
Suppression 5‰ (fresh weight) [%]	98	59.1	14.0 - 103.0	3	18.7	0.47

3.2.3. Conclusion

Michel et al. [42] concluded that NIR spectroscopy was a capable method to predict various chemical and biological parameters using PLS regression. They believe NIR spectroscopy to be capable of monitoring compost quality.

3.3. Cluster Analysis (CA)

3.3.1. Objective of the study

The objective of the study by Ros et al. [27] was to find out the long-term effects of composts on soil microbial communities. Different types of compost were applied over a period of 12 years. DNA was extracted by Ros et al. [27] from differently treated soils. The microbial community was described by polymerase chain reaction coupled with denaturing gradient gel electrophoresis (PCR-DGGE). They used multivariate data analysis to show the differences or similarities of microbial communities using DGGE data.

3.3.2. Method of evaluation and results

A polymerase chain reaction coupled with denaturing gradient gel electrophoresis (PCR-DGGE) was performed to characterize the microbial community. In Fig. 4 a DGGE fingerprint is shown. For the interpretation of such fingerprints statistical tools are necessary. DGGE data were converted into a binary system for cluster analysis (Fig. 4). As mentioned above, cluster analysis visualises the similarity between the samples in a dendrogram.

Ros et al. [27] show the cluster analysis of the DGGE profiles of 16S rDNA from the whole bacterial community. The cluster analysis illustrates the segregation of two soil groups. The clusters are caused by two different amendments. One cluster comprises the soil with compost and nitrogen application, the second cluster represents the soil with amendment of different composts (compost + nitrogen as mineral fertiliser).

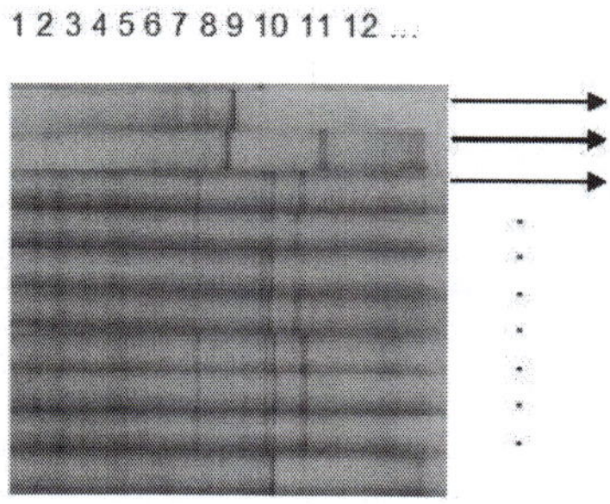

	1	2	3	4	5	6	7	8	9	10	11	12
80	0	0	0	0	0	0	0	1	1	0	0	0
GC	0	0	0	0	0	0	0	1	1	0	0	1
OWC	0	1	1	0	0	1	1	1	1	0	0	0
MC	0	1	1	0	0	1	1	1	1	0	0	0
SSC	0	1	1	0	0	1	1	1	1	0	0	0
Control	0	1	1	0	0	1	1	1	1	0	0	0
GC+80	0	1	0	0	0	1	1	1	1	0	1	1
OWC+80	0	1	1	0	0	1	1	0	1	0	0	0
MC+80	0	1	1	0	1	1	1	1	1	0	0	0
SSC+80	0	1	1	0	0	1	1	1	1	0	0	0

FIGURE 4. DGGE fingerprint and an example of a binary DGGE data matrix

3.3.3. Conclusion

Ros et al. [27] concluded that the differences between soils with compost with additional nitrogen fertiliser, and the second cluster comprising compost, control and mineral fertiliser soils are stronger than the influence of the different compost types. Furthermore they hypothesised that a certain microbial community inherent to the different composts is irrelevant after 12 years of compost application. Based on the cluster analyses of the PCR-DGGE data, they concluded that the combined application of compost and nitrogen affected soil properties regarding microbial communities much more.

3.4. Soft Independent Modelling Of Class Analogy (SIMCA)

3.4.1. Objective of the study

Malley et al. [8] used a portable near infrared (NIR) spectrometer to investigate changes of biogenic waste materials during composting. The idea of this study was to observe the composting process continuously in an easy and inexpensive way using NIR spectroscopy.

3.4.2. Method of evaluation and results

First of all many spectra were collected by Malley et al. [8]. The interpretation of spectral data requires experience in spectral interpretation. To provide rapid interpretation of the measured infrared spectra Malley et al. [8] applied the classification method SIMCA. The SIMCA model allows the assignment of a new sample to a defined class. A SIMCA model is always based on the PCAs of the various defined classes. Malley et al. [8] defined 3 different classes: raw manure (M), stockpiled manure (S) and manure compost (C). In the study 2 years of composting were observed (2000 and 2001).Figure 2 by Malley et al. [8] shows the scores plot of the PCA based on the spectral data of the three different classes in the year 2001. The PCA demonstrates a clear grouping of the 3 classes manure, stockpiled manure and manure compost.

Malley et al. [8] illustrated the results of the SIMCA by means of a Coomans plot. In figure 3 by Malley et al. [8] they show the Coomans plot for the investigations of 2001. The vertical and horizontal lines in the Coomans plot mark the 5 % level of significance. That means that 95 % of the samples that truly belong to this group are found within the line. Due to the fact that compost lies on the opposite side of the vertical line from the raw and stockpiled samples Malley et al. [8] concluded that compost is significantly different from the other two classes. The groups of raw manure and stockpiled manure are overlapping. Thus Malley et al. [8] concluded that they did not differ significantly. Nevertheless some raw samples were different. With these results Malley et al. [8] demonstrated that spectroscopic data and multivariate data analysis, especially SIMCA provides a sensitive analysis to differentiate between the products of stockpiles and compost.

3.4.3. Conclusion

Malley et al. [8] concluded that NIR spectroscopy and the multivariate data analysis method SIMCA can be a rapid, inexpensive method for assessing a composting process.

4. CRITICAL DISCUSSION OF MULTIVARIATE STATISTICAL METHODS

In fact there are some statistical restrictions, which cannot be solved easily. The simple situation starts with the general linear model. This model usually has a character variable y depending on one or more predictor variables $x_1, x_2, ..., x_k$:

In case of cross-classified two-way analysis of variance (equal subclass numbers):

$$y_{ijk} = \mu + a_i + b_j + w_{ij} + e_{ijk}, \ (i = 1, \quad , a; j = 1, \dots, b; k = 1, \dots, n) \quad (1)$$

μ is the general mean, a_i are the main effects of factor A, b_j are the main effects of factor B, w_{ij} are the interactions between A_i and B_j, e_{ijk} are the random error terms.

In case of multiple linear regression:

$$y_j = \beta_0 + \beta_1 x_{1j} + \beta_2 x_{2j} + \dots + \beta_k x_{kj} + e_j, \ (j = 1, \dots, n), \quad (2)$$

y_j is the j-th value of y depending on the j-th values $x_{1j}, ... x_{kj}$;

e_j are error terms with $E(e_j) = 0$, $var(e_j) = \sigma^2$ (for all j), $cov(e_{j'}, e_j) = 0$ for $j' \neq j$
The simple case assumes a linear dependency. The statistical parameters (the model coefficients) of the model can be estimated, y can be estimated for given values $x_1, ... x_k$. Assuming that the e_j are normally distributed, confidence intervals can be calculated for each model coefficient and finally tests of hypotheses about the model coefficients can be performed. By this procedure each variable can be tested whether its influence on the variable y is significantly different from 0 or not. The type I and type II error can be stated. Furthermore optimal designs for the experiments and surveys can be calculated [89]. Several assumptions are typically made regarding the distribution of the populations and regarding homoscedasticity. Furthermore the problem of extreme values and outliers respectively is critical, especially in environmental measurements. Increasing the number of regressors and factors respectively also increases the error terms.

For some univariate models robust and powerful alternatives regarding the distribution assumptions and regarding homoscedasticity [90-92] already exist. In the case of cross classification there is still no satisfying, powerful alternative. Many multiple regressors methods (multiple regression models, logistic regression models, discriminant analysis, cross classification models) need independent variables.

In chemometrics some of these problems are highly relevant. Usually the number of regressor variables exceeds the number of samples, which excludes most of the common oligovariate models. Many of the regressor variables are highly collinear. Due to these reasons dimension reduction methods are used such as correspondence analysis or factor analysis. The new factors in the latter are strictly independent from one another and can therefore be used in conventional models. There are several possibilities to extract these factors, like Principal Components or Maximum Likelihood. A possibility to model discrete variables is the classification by means of cluster analysis. These clusters can be tested later by contingency tables. Both steps (factor analysis and cluster analysis) lead to descriptive variables of the data set. Just as all descriptive methods in statistics they do not serve as tests against hypothesis of pure chance. There is no risk assessment of the results. Testing of the new descriptive variables implies the understanding of these new variables. By loading the original variables onto the new variables sometimes the interpretation can be done easily. Then models with these variables can be established (PCR or PLS-R) with several quality parameters (e.g. correlation coefficient). A test of significance for the cross-validated r^2 was performed by Wakeling and Morris [93]. In this paper critical values of r^2 occurring just by chance alone are tabulated for one to three dimensional models at a significance level of 5 % based on Monte Carlo simulations. A comparable method was used by Stahle and Wold [94] to develop a polynomial approximation of the test statistic for the two-class problem and the number of objects, the number of variables, the percentage variance explained by the first component in X and the percentage of missing values.

$$cvd/sd = \sqrt{PRESS/RSS} \qquad (3)$$

cvd: cross-validated deviances

sd: standard deviation

PRESS: prediction error sum of squares

RSS: residual sum of squares

Unfortunately the definition of hypothesis regarding the regression coefficients still refers to the new components and provides no results regarding the original variables. There is no statistical possibility to prove whether the extraction method is optimal. Other methods of dimension reduction are already in use (e.g. Boosting, Random forest). Robust alternatives for PLS-R are also available [95].

As long as there are no satisfying testing routines, the results of the presented multivariate methods have to be interpreted very carefully. There is an inherent risk of over-interpretation, especially when using descriptive methods such as PCA or cluster analysis. There is no definition of the error probability of the results. That means whatever interpretation of the picture is done, it could be just pure coincidence and there is no information about the risk. The only possibility to overcome these problems would be to analyse a large number of samples and in case of regression models to validate these models.

5. SUMMARY

In waste management research and practice often huge data sets for statistical evaluation are required to verify the findings. This request concerns both the natural scientific and the logistic field of waste management. Huge data sets can be generated on the one hand by vast numbers of investigated parameters and samples and on the other hand by modern analytical methods such as spectroscopic, chromatographic methods or thermal analysis.

Multivariate data analysis can help to explore data structures of the investigated samples. Another advantage is that the results can be displayed graphically. Furthermore, validated models can serve as adequate evaluation tools for practical application. Different software types are offered to develop such evaluation tools.

In this study the most important multivariate data analysis methods applied in waste management were described in detail and documented by a literature review. It could be demonstrated that Principal Component Analysis (PCA) and Partial Least Square Regression (PLS-R) are the most applied methods in waste management. PCA was used to find hidden data structures, groupings and interrelationships of data. In most cases PLS-R was applied to predict parameters using new analytical instruments that allow faster and cheaper analyses.

In general it can be stated that multivariate data analysis was successfully applied in all experiments. Several authors compared different multivariate methods to determine which one provided the best results. Depending on the data set and the question to be answered the appropriate method must be identified.

REFERENCES

1. R Zbytniewski, B Buszewski, 2005Characterization of natural organic matter (NOM) derived from sewage sludge compost. Part 2: Multivariate techniques in the study of compost maturation. Bioresour. Technol. 96479484

2. R Albrecht, R Joffre, R Gros, Le Petit J, Terrom G, Périssol C (2008Efficiency of near-infrared reflectance spectroscopy to assess and predict the stage of transformation of organic matter in the composting processBioresour. Technol. 99448455

3. P Campitelli, , S Ceppi, (2008) Chemical, physical and biological compost and vermicompost characterization: A chemometric study. Chemometrics Intell. Lab. Syst. 90 6471 .

4. M. V Gil, L. F Calvo, D Blanco, M. E Sánchez, 2008Assessing the agronomic and environmental effects of the application of cattle manure compost on soil by multivariate methodsBioresour. Technol. 9957635772

5. M Grigatti, L Cavani, C Ciavatta, 2007A multivariate approach to the study of the composting process by means of analytical electrofocusingWaste Manage. 2710721082

6. M Hansson, A Nordberg, B Mathisen, 2003On-line NIR monitoring during anaerobic treatment of municipal solid waste.Water Sci. Technol. 48913

7. LaMontagne MGMichel Jr FC, Holden PA, Reddy CA (2002Evaluation of extraction and purification methods for obtaining PCR-amplifiable DNA from compost for microbial community analysis.J. Microbiol. Meth. 49255264

8. D. F Malley, C Mcclure, P. D Martin, K Buckley, W. P Mccaughey, 2005Compositional analysis of cattle manure during composting using a field-portable near-infrared spectrometerCommunications in Soil Science and Plant Analysis36455475

9. J Nicolas, A. C Romain, V Wiertz, J Maternova, P Andre, 2000Using the classification model of an electronic nose to assign unknown malodours to environmental sources and to monitor them continuouslySensors and Actuators, B: Chemical 69366371

10. P Planquart, G Bonin, A Prone, C Massiani, 1999Distribution, movement and plant availability of trace metals in soils amended with sewage sludge composts: Application to low metal loadingsSci. Total Environ. 241161179

11. A. C Romain, D Godefroid, M Kuske, J Nicolas, 2005Monitoring the exhaust air of a compost pile as a process variable with an e-noseSensors and Actuators, B: Chemical 1062935

12. E Smidt, K Meissl, M Schwanninger, P Lechner, 2008Classification of waste materials using Fourier transform infrared spectroscopy and soft independent modeling of class analogyWaste Manage. 2816991710

13. A. J Termorshuizen, E Van Rijn, D. J Van Der Gaag, C Alabouvette, Y Chen, J Lagerlöf, et al2006Suppressiveness of 18 composts against 7 pathosystems: Variability in pathogen response. Soil Biol. Biochem. 3824612477

14. K Böhm, E Smidt, J Tintner, 2011Modelled on Nature- Biological Processes in Waste Management. In: Kumar S, editor. Integrated Waste Management. Rijeka, Croatia: InTech. 153178

15. E Smidt, J Tintner, K Böhm, E Binner, 2011Transformation of biogenic waste materials through anaerobic digestion and subsequent composting of the residues- A case study. Dyn. Soil Dyn. Plant 56369

16. E Smidt, K Meissl, J Tintner, E Binner, 2008Influence of Input Materials and Composting Operation on Humification of Organic Matter. In: Hao X, editor. Dyn. Soil Dyn. Plant. Special 15059

17. E Smidt, K Böhm, M Schwanninger, 2011The Application of FT-IR Spectroscopy in Waste Management. In: G.S. N, editor. Fourier Transforms- New Analytical Approaches and FTIR Strategies. Rijeka, Croatia: InTech. 405430

18. G Bianchi, G Celeste, M Palmiotto, E Davoli, 2010Source identification of odours and VOCs from a composting plant by multivariate analysis of trace volatile organic compoundsChemical Engineering Transactions23279284

19. F Biasioli, E Aprea, F Gasperi, T. D Mark, 2009Measuring odour emission and biofilter efficiency in composting plants by proton transfer reaction-mass spectrometry.Water Sci. Technol. 5912631269

20. C Tognetti, M. J Mazzarino, F Laos, 2011Comprehensive quality assessment of municipal organic waste composts produced by different preparation methodsWaste Manage. 3111461152

21. A Vergnoux, M Guiliano, Le Dréau Y, Kister J, Dupuy N, Doumenq 2009Monitoring of the evolution of an industrial compost and prediction of some compost properties by NIR spectroscopySci. Total Environ. 407: 2390-2403.

22. X. S He, B. D Xi, Y. H Jiang, L. S He, D Li, H. W Pan, et al2012Structural transformation study of water-extractable organic matter during the industrial composting of cattle manureMicrochem. J. in press: DOI:j.microc.2012.1006.1004.

23. P Geladi, B Sethson, m J Nystro, T Lillhonga, T Lestander, J Burger, 2004Chemometrics in spectroscopy: Part 2. Examples. Spectrochimica Acta- Part B Atomic Spectroscopy 5913471357

24. M. A Bustamante, F Suarez-estrella, C Torrecillas, C Paredes, R Moral, J Moreno, 2010Use of chemometrics in the chemical and microbiological characterization of composts from agroindustrial wastesBioresour. Technol. 10140684074

25. I. H Franke-whittle, B. A Knapp, J Fuchs, R Kaufmann, H Insam, 2009Application of COMPOCHIP Microarray to Investigate the Bacterial Communities of Different CompostsMicrobial Ecol. 57510521

26. G Innerebner, B Knapp, T Vasara, M Romantschuk, H Insam, 2006Traceability of ammonia-oxidizing bacteria in compost-treated soilsSoil Biol. Biochem. 3810921100

27. M Ros, S Klammer, B Knapp, K Aichberger, H Insam, 2006Long-term effects of compost amendment of soil on functional and structural diversity and microbial activitySoil Use Manage. 22209218

28. M Ros, J. A Pascual, C Garcia, M. T Hernandez, H Insam, 2006Hydrolase activities, microbial biomass and bacterial community in a soil after long-term amendment with different compostsSoil Biol. Biochem. 3834433452

29. W Sun, G. H Huang, G. M Zeng, X. S Qin, X. L Sun, 2009A Stepwise-cluster, microbial biomass inference model in food waste composting. Waste Manage. 2929562968

30. W Sun, G. H Huang, G. M Zeng, X. S Qin, H Yu, 2011Quantitative effects of composting state variables on C/N ratio through GA-aided multivariate analysisSci. Total Environ. 40912431254

31. Y Xiao, G. M Zeng, Z. H Yang, Y. H Ma, C Huang, W. J Shi, et al2011Effects of Continuous Thermophilic Composting (CTC) on Bacterial Community in the Active Composting ProcessMicrobial Ecol. 62599608

32. M Chikae, R Ikeda, K Kerman, Y Morita, E Tamiya, 2006Estimation of maturity of compost from food wastes and agro-residues by multiple regression analysisBioresour. Technol. 9719791985

33. K Suehara, T Yano, 2004Bioprocess monitoring using near-infrared spectroscopy.Advances in biochemical engineering/biotechnology90173198

34. P Capriel, T Ebertseder, L Popp, R Gutser, 1999IR-spectroscopy: A rapid method to estimate the nitrogen effect and relevant parameters of biocompostsIR-Spektroskopie: Eine Schnellmethode zur Prognose der N-Wirkung und relevanten Parameter von Biokomposten 162149153

35. T Fujiwara, K Murakami, 2007Application of near infrared spectroscopy for estimating available nitrogen in poultry manure compost. Soil Science and Plant Nutrition53102107

36. G Huang, L Han, X Liu, 2007Rapid estimation of the composition of animal manure compost by near infrared reflectance spectroscopyJ. Near Infrared Spectrosc. 15387394

37. G Huang, L Han, Z Yang, X Wang, 2008Evaluation of the nutrient metal content in Chinese animal manure compost using near infrared spectroscopy (NIRS)Bioresour. Technol. 9981648169

38. H. J Ko, H. L Choi, H. S Park, H. W Lee, 2004Prediction of heavy metal content in compost using near-infrared reflectance spectroscopyAsian-Australasian Journal of Animal Sciences 1717361740

39. B Ludwig, G Schmilewski, T Terhoeven-urselmans, 2006Use of near infrared spectroscopy to predict chemical parameters and phytotoxicity of peats and growing mediaScientia Horticulturae1098691

40. K Meissl, E Smidt, M Schwanninger, 2007Prediction of humic acid content and respiration activity of biogenic waste by means of Fourier transform infrared (FTIR) spectra and partial least squares regression (PLS-R) modelsTalanta72791799

41. K Meissl, E Smidt, M Schwanninger, J Tintner, 2008Determination of humic acids content in composts by means of near- And mid-infrared spectroscopy and partial least squares regression models.Appl. Spectrosc. 62873880

42. K Michel, C Bruns, T Terhoeven-urselmans, B Kleikamp, B Ludwig, 2006Determination of chemical and biological properties of composts using near infrared spectroscopyJ. Near Infrared Spectrosc. 14251259

43. S Morimoto, W. F Mcclure, B Crowell, D. L Stanfield, 2003Near infrared technology for precision environmental measurements: Part 2. Determination of carbon in green grass tissueJ. Near Infrared Spectrosc. 11257267

44. F Zvomuya, B. L Helgason, F. J Larney, H. H Janzen, O. O Akinremi, B. M Olson, 2006Predicting phosphorus availability from soil-applied composted and non-composted cattle feedlot manure.Journal of Environmental Quality35928937

45. L Galvez-sola, R Moral, M. D Perez-murcia, A Perez-espinosa, M. A Bustamante, E Martinez-sabater, et al2010The potential of near infrared reflectance spectroscopy (NIRS) for the estimation of agroindustrial compost qualitySci. Total Environ. 40814141421

46. A Mohajer, A Tremier, S Barrington, J Martinez, C Teglia, M Carone, 2009Microbial oxygen uptake in sludge as influenced by compost physical parametersWaste Manage. 2922572264

47. Sharma HSSKilpatrick M, Lyons G, Sturgeon S, Archer J, Moore S, et al. (2005Visible and near-infrared calibrations for quality assessment of fresh phase I and II mushroom (Agaricus bisporus) compostAppl. Spectrosc. 5913991405

48. K. I Suehara, Y Nakano, T Yano, 2001Simultaneous measurement of carbon and nitrogen content of compost using near infrared spectroscopyJ. Near Infrared Spectrosc. 93541

49. L Galvez-sola, J Morales, A. M Mayoral, F. C Marhuenda-egea, E Martinez-sabater, M. D Perez-murcia, et al2010Estimation of phosphorus content and dynamics during composting: Use of near infrared spectroscopyChemosphere781321

50. H Ecke, 2003Sequestration of metals in carbonated municipal solid waste incineration (MSWI) fly ashWaste Manage. 23631640

51. P Mostbauer, P Lechner, K Meissl, 2007Langzeitverhalten von MVA-Reststoffen- Evaluierung von Testmethoden. Vienna: facultas.wuv. 158 p.

52. B Scaglia, F Adani, 2008An index for quantifying the aerobic reactivity of municipal solid wastes and derived waste productsScience of the Total Environment394183191

53. E Smidt, K Böhm, J Tintner, 2010Application of various statistical methods to evaluate thermo-analytical data of mechanically-biologically treated municipal solid waste.Thermochim. Acta 5019197

54. R Abouelwafa, Ait Baddi G, Souabi S, Winterton P, Cegarra J, Hafidi M (2008Aerobic biodegradation of sludge from the effluent of a vegetable oil processing plant mixed with household waste: Physical-chemical, microbiological, and spectroscopic analysisBioresour. Technol. 9985718577

55. R Abouelwafa, S Amir, S Souabi, P Winterton, V Ndira, J. C Revel, et al2008The fulvic acid fraction as it changes in the mature phase of vegetable oil-mill sludge and domestic waste compostingBioresour. Technol. 9961126118

56. X. S He, B. D Xi, Z. M Wei, X. J Guo, M. X Li, D An, et al2011Spectroscopic characterization of water extractable organic matter during composting of municipal solid wasteChemosphere82541548

57. K Böhm, E Smidt, E Binner, M Schwanninger, J Tintner, P Lechner, 2010Determination of MBT-waste reactivity- An infrared spectroscopic and multivariate statistical approach to identify and avoid failures of biological tests. Waste Manage. 30583590

58. F Biasioli, F Gasperi, G Odorizzi, E Aprea, D Mott, F Marini, et al2004PTR-MS monitoring of odour emissions from composting plantsInt. J. Mass Spectrom. 239103109

59. E De Rosa, D Rubel, M Tudino, A Viale, R. J Lombardo, 1996The leachate composition of an old waste dump connected to groundwater: Influence of the reclamation worksEnvironmental Monitoring and Assessment40239252

60. E Durmusoglu, C Yilmaz, 2006Evaluation and temporal variation of raw and pre-treated leachate quality from an active solid waste landfillWaterAir, and Soil Pollution 171359382

61. K Kylefors, 2003Evaluation of leachate composition by multivariate data analysis (MVDA).J. Environ. Manage. 68367376

62. E. V Mikhailov, O. V Tupicina, D. E Bykov, K. L Chertes, O. Y Rodionova, A. L Pomerantsev, 2007Ecological assessment of landfills with multivariate analysis- A feasibility study. Chemometrics Intell. Lab. Syst. 87147154

63. J Olivero-verbel, C Padilla-bottet, O De La Rosa, 2008Relationships between physicochemical parameters and the toxicity of leachates from a municipal solid waste landfillEcotox. Environ. Safe. 70294299

64. E Smidt, J Tintner, K Meissl, 2007New approaches in landfill assessment. In:Velinni AA, editor. Landfill Research Trends. New York: Nova Science Publisher, Inc. 291

65. S. K Srivastava, A. L Ramanathan, 2008Geochemical assessment of groundwater quality in vicinity of Bhalswa landfill, Delhi, India, using graphical and multivariate statistical methodsEnvironmental Geology5315091528

66. E Smidt, K Böhm, J Tintner, 2011Monitoring and assessment of landfills using simultaneous thermal analysisSustain. Environ. Res. 21247252

67. S Diener, L Andreas, I Herrmann, H Ecke, A Lagerkvist, 2010Accelerated carbonation of steel slags in a landfill cover constructionWaste Manage. 30132139

68. M. V Pablos, C Fernandez, M. D Babin, J. M Navas, G Carbonell, F Martini, et al2009Use of a novel battery of bioassays for the biological characterisation of hazardous wastesEcotox. Environ. Safe. 7215941600

69. J Tintner, B Klug, 2011Can vegetation indicate landfill cover features? Flora 206559566

70. M Van Praagh, K. M Persson, P Karlsson, 2009Potential environmental impacts of using refuse derived material for landfill cappingWaste Manage. Res. 27471488

71. H Ecke, A Bergman, A Lagerkvist, 1998Multivariate data analysis (MVDA) in landfill researchJournal of Solid Waste Technology and Management253339

72. G Jean, J. F Fruget, 1994Comparison of ecotoxicological and physico-chemical data by use of multivariate analyses and graphical displaysChemosphere2822492267

73. El-M. A Sheikh, Y. M Al-sodany, E. M Eid, K. H Shaltout, 2012Ten years primary succession on a newly created landfill at a lagoon of the Mediterranean Sea (Lake Burullus RAMSAR site)Flora207459468

74. K. D Kim, E. J Lee, K. H Cho, 2004The plant community of Nanjido, a representative nonsanitary landfill in South Korea: Implications for restoration alternativesWaterAir, and Soil Pollution 154167185

75. M Tesar, R Prantl, P Lechner, 2007Application of FT-IR for assessment of the biological stability of landfilled municipal solid waste (MSW) during in situ aerationJournal of Environmental Monitoring9110118

76. E Smidt, K Meissl, J Tintner, F Ottner, 2010Interferences of carbonate quantification in municipal solid waste incinerator bottom ash: Evaluation of different methodsEnviron. Chem. Lett. 8217222

77. M Huber-humer, J Tintner, K Bohm, P Lechner, 2011Scrutinizing compost properties and their impact on methane oxidation efficiencyWaste Manage. 31871883

78. E Smidt, K Böhm, J Tintner, 2010Evaluation of old landfills- A thermoanalytical and spectroscopic approach. J. Environ. Monitor. 13362369

79. T. A Lawrence, G. P Boutwell, 1990Predicting stratigraphy at landfill sites using electromagnetics. In: Landva A, Knowles DG, editors. Geotechnics of Waste Fills-Theory and Practice; 10-13. September 1989; Pittsburgh, PA, USA: ASTM Special Technical Publication; 3140

80. V Moreno-santini, I Mansilla-rivera, O García-rodríguez, C. J Rodríguez-sierra, 2012A Pilot Study Determining Hair Arsenic and Lead Levels in Residents of a Community Established on a Former Landfill in Puerto RicoB. Environ. Contam. Tox.: 15

81. L Dahlén, S Vukicevic, J. E Meijer, A Lagerkvist, 2007Comparison of different collection systems for sorted household waste in SwedenWaste Manage. 2712981305

82. C Lin, Wu EMY, Lee CN, Kuo SL (2011Multivariate Statistical Factor and Cluster Analyses for Selecting Food Waste Optimal Recycling MethodsEnviron. Eng. Sci. 28349356

83. Y. W Cheng, F. C Sung, Y Yang, Y. H Lo, Y. T Chung, K. C Li, 2009Medical waste production at hospitals and associated factorsWaste Manage. 29440444

84. R Noori, M. A Abdoli, M. J Ghazizade, R Samieifard, 2009Comparison of Neural Network and Principal Component-Regression Analysis to Predict the Solid Waste Generation in TehranIran. J. Public Health 387484

85. K Esbensen, 2002Multivariate Data Analysisin practice. Esbjerg: Alborg University. 598 p.

86. W. F Velicer, D. N Jackson, 1990Component Analysis versus Common Factor Analysis: Some issues in Selecting an Appropriate Procedure.Multivar. Behav. Res. 25128

87. ter Braak CJEVerdonschot PEM (1995Canonical correspondence analysis and related multivariate methods in aquatic ecologyAquat. Sci. 57255289

88. R. G Brereton, 2002Chemometrics: Data analysis for the laboratory and chemical plant.Chichester, England: John Wiley & Sons Ltd. 489 p.

89. D Rasch, L. R Verdooren, J. I Gowers, 1999Fundamentals in Design and Analysis of Experiments and Surveys. Munich, Vienna: Oldenbourg Wissenschaftsverlag GmbH. 253 p.

90. K Moder, 2007How to keep the Type I Error Rate in ANOVA if Variances are Heteroscedastic. Austrian Journal of Statistics 36179188

91. K Moder, 2010Alternatives to F-Test in One Way ANOVA in case of heterogeneity of variances (a simulation study)Psychological Test and Assessment Modeling52343353

92. D Rasch, K Kubinger, K Moder, 2011The two-sample t test: Pre-testing its assumptions does not pay offStat. Pap. 52219231

93. I. N Wakeling, J Morris, 1993A test of significance for partial least squares regressionJ. Chemometr. 7291304

94. L Stahle, S Wold, 1987Partial least squares analysis with cross-validation for the two-class problem: a Monte Carlo study. J. Chemometr. 1185196

95. I. N Wakeling, Macfie HJH (1992A robust PLS procedure. J. Chemometr. 6189198

6

CHAPTER

Estimation of Reference Evapotranspiration Using Multivariate Fractional Polynomial, Bayesian Regression, and Robust Regression Models in Three Arid Environments

Mojtaba Khoshravesh[1] • Mohammad Ali Gholami Sefidkouhi[1] • Mohammad Valipour[2]

[1]*Department of Water Engineering, Sari Agricultural Sciences and Natural Resources University, Sari, Iran*

[2]*Young Researchers and Elite Club, Kermanshah Branch, Islamic Azad University, Kermanshah, Iran*

ABSTRACT

The proper evaluation of evapotranspiration is essential in food security investigation, farm management, pollution detection, irrigation scheduling, nutrient flows, carbon balance as well as hydrologic modeling, especially in arid environments. To achieve sustainable development and to ensure water supply, especially in arid environments, irrigation experts need tools to estimate reference evapotranspiration on a large scale. In this study, the monthly reference evapotranspiration was estimated by three different regression models including the multivariate fractional polynomial (MFP), robust regression, and Bayesian regression in Ardestan, Esfahan, and Kashan. The results were compared with Food and Agriculture Organization (FAO)-Penman–Monteith (FAO-PM) to select the best model. The results show that at a monthly scale, all models provided a closer agreement with the calculated values for FAO-PM ($R^2 > 0.95$ and RMSE < 12.07 mm month^{-1}). However, the MFP model gives better estimates than the other two models for estimating reference evapotranspiration at all stations.

Keywords: Evapotranspiration Iran Multivariate fractional polynomial Bayesian regression Robust regression

INTRODUCTION

Evapotranspiration (ET) is one of the major elements of the hydrologic cycle, and its accurate predictions is of paramount importance for many investigations such as irrigation system design and management, hydrologic water balance, crop yield simulation, irrigation scheduling, drainage studies, agricultural and forest meteorology, and water resources planning and management (Banihabib et al. 2012; Kumar et al.2002; Valipour 2014c, d, e, f, g, h).

The most important parameters that affect on variations of evapotranspiration are air temperature, humidity, wind speed, and sun radiation (Pejic et al. 2015; Valipour 2014a,i, j, k; Wrachien and Mambretti 2015; Yannopoulos et al. 2015). At first, there is a distinction between reference and actual evapotranspiration. Reference evapotranspiration corresponds to the amount of water that would return to the atmosphere in case of moisture redundancy, for a reference crop [e.g., alfalfa (ET_r) or grass (ET_0)], from a standard surface and to then apply an appropriate empirical crop coefficient, which accounts for the difference between the standard surface and crop ET (Tzimopoulos et al. 2007; Valipour 2015a, b, c; Valipour et al. 2012).

ET_0 refers to the water emitted from a unit ground area covered with a reference plant, unstressed and healthy as well as with ample water supply (Walter et al. 2002). The ET_0 is used to quantify evapotranspirative demand within a zone and to forecast crop evapotranspiration when the ET_0 is multiplied by a crop coefficient (K_c) to determine for differences between the grass and plant evapotranspiration (Allen et al. 1998; Schuch and Burger 1997).

Measurements of pan evaporation are spatiotemporally limited, especially in developing countries. In locations where pan evaporation measurements are sparse, theoretical forecasts can be used to estimate it from other existing data, which depend strongly on the local conditions. Evapotranspiration is a function of wind speed, temperature, and humidity. Solar radiation is also a significant parameter for evapotranspiration. All of these factors do not act independently. Wind varies the humidity, and if the humidity is low, the evapotranspiration rate is higher. At the same time, air temperature affects surface temperature and humidity. These parameters all contribute to evaporation with temperature likely being the dominant factor influencing evapotranspiration (Rahimikhoob 2009).

Where pan evaporation data are not available, evapotranspiration can be characterized by evapotranspiration models using weather variables. Among them, the Penman model is most frequently used for evapotranspiration prediction. The Penman model requires five climatic parameters: temperature, relative humidity, wind, saturation vapor pressure, and net radiation. It also uses complicated unit conversions and lengthy computations (Wu 1997).

Many researchers have evaluated the reliability of the Penman–Monteith (PM) model for predicting ET_0 (McNaughton and Jarvis 1984; Allen 1986; Allen et al. 1989; De Souza and Yoder 1994; Chiew et al. 1995). Jensen et al. (1990) analyzed the performance of 20 different models against lysimeter measured ET for 11 stations located in various climatic regions around the world. The PM model ranked as the best method for all climatic situations. However, the ranking of other models varied, depending on their local calibration and conditions.

Abtew and Obeysekera (1995) and Abtew (1996) found that the Penman–Monteith model was well suited to predict evapotranspiration from cattails (Typha domingensis), mixed marsh vegetation, and an open water/algae system; however, that calibrated simpler radiation-based methods also provided reasonable forecasts.

Potential evapotranspiration (PET), rather than actual evapotranspiration (AET), is a common input for empirical methods because it offers an upper limit to evapotranspirative water losses. PET is a function of vapor pressure gradient, available energy, and vegetation type (Douglas et al. 2009). In predicting PET, a clear definition of the "best" method for calculation is not evident and the model choice is often subjective. Verstraeten et al. (2008) presented a comprehensive review of the scientific articles on models for evaluating PET and mentioned that the selection of one model from the many is primarily dependent on the goals of the investigation and the type of available data. For example, Weib and Menzel (2008) compared the Priestley–Taylor (PT) method, two methods based on the Penman–Monteith (PM) equation and the Hargreaves method, a temperature-based method for predicting PET in a global-scale hydrologic method.

However, more study is needed on irrigation management to establish the appropriate model to be applied for evaluating crop evapotranspiration (ET), to avoid the excess or deficit water application, waterlogging, and soil salinity (Eslamian et al. 2009). Operational software tools in the domains specified above require the evaluation of $ET_0$and the procedures to do that have been repeatedly implemented in software applications. This is one of the reasons that why there is an increasing demand for modular methods in model development (Jones et al. 2001). Reference evapotranspiration can be measured by lysimeters; however, this approach is very expensive and not easy to use (Valipour 2014l, m, 2015e). Many authors have attempted to estimate the evaporation through the indirect models using the weather parameters; however, some of these methods require the data, which cannot easily be measured (Rosenberry et al. 2007).

Today, applications of modeling and regression methods have been widely spread in different fields of the science. These methods were applied for increasing the quality of forecasts using the sample data sets from experiments and different investigations. Appropriate use of these methods is very useful for resolve of uncertainties and results of predictions, when obtain of information is difficult and sometimes impossible.

The most frequently used statistical methods are known as frequentist (or classical) methods. These methods assume that unknown parameters are the fixed constants, and they define probability using limiting relative frequencies. Bayesian methods offer an alternative approach; they treat the parameters as random variables and define probability as "degrees of belief" (i.e., the probability of an event is the degree to believe the event is true). The term "Bayesian" comes from the prevalent usage of Bayes' theorem, which was named after the Reverend Thomas Bayes, an eighteenth-century Presbyterian minister (Rathnayake 2010).

Bayesian analysis is a mode of inductive reasoning that has been applied in a lot of scientific disciplines. A distinctive feature of the Bayesian approach is that it permits the researcher to use sample (data) and prior (expert judgment) information in a logically consistent manner in making inferences. This is done by using Bayes' theorem to produce a "post-data" or posterior distribution for the model variables. Using Bayes' theorem, prior (or initial) values are transformed to post-data views. This transformation can be viewed as a learning process. The posterior distribution is calculated by the variances of the prior and sample information. If the variance of the prior information is smaller than the variance of the sampling information, then a higher weight is assigned to the prior information. On the other hand, if the variance of the sample information is smaller than the variance of the prior information, then a higher weight is assigned to the sample information causing the posterior forecast to be closer to the sample information (Chulani et al. 1998; Singh 2014).

Since their introduction by Royston and Altman (1994), the regression methods based on fractional polynomial (FP) transformation of continuous predictor(s) have found gradual acceptance as a useful technique of analysis which retains such predictors as continuous in the model. The robust regression using functions related to the least squares has been the subject of intense research (Royston and Sauerbrei 2007). Demoster et al. (1980) discussed statistical properties of the predictions under the assumption that the observation errors are independent normal. Coleman et al. (1980) developed a high-quality set of routines to calculate robust predictions for eight weight functions.

The superiority of regression models to artificial intelligence techniques such as ANFIS, ANN, and SVM is that regression models have the less complexity. In addition, these models give us the formula or relationship and have an acceptable accuracy. Although application of nonlinear regression and Bayesian approach has been surveyed in the various fields of hydrology and water management such as irrigation canal demands (Ticlavilca et al. 2013), low flow indices (Joshi et al. 2013), land surface temperature (Ghosh and Joshi 2014), catchment modeling (Marshall et al. 2007), nutrient loading (Vigiak and Bende-Mich 2013; Wellen et al. 2012), heat flux (Ershadi et al. 2013; Yao et al. 2014), solar radiation (Iizumi et al. 2012), and streamflow modeling (Block and Rajagopalan 2009; Liang et al. 2013), there are few studies about application of them in estimation of ET (Bachour et al. 2014; Izadifar and Elshorbagy 2010; Tabari et al. 2012; Zhu et al. 2013). However, these studies focus on some limited methods and/or use low recorded data. Furthermore, comparison of the nonlinear regression methods and Bayesian approach has not been investigated in arid regions (Valipour 2013a, b). In the present paper, the attempts are made to apply MFP, Bayesian regression, and robust regression models to estimate reference evapotranspiration values, in three arid regions of Iran in which water crises are considerable (Mahdizadeh Khasraghi et al. 2015; Valipour et al. 2015), using temperature data.

MATERIALS AND METHODS

Fractional polynomials

Simple power transformations of a covariate have long been used informally in data analysis in cases when nonlinearity was suspected. General power models were proposed by Box and Tidwell (1962). Royston and Altman (1994) formalized the simple power models and called them fractional polynomials of degree 1 (FP1), and extended them to FPs of higher degree. Ignoring the intercept term (β_0), a polynomial of degree m in a single covariate X may be written:

$$\beta_1 X^1 + \beta_2 X^2 + \cdots + \beta_m X^m \tag{1}$$

In the generalization to an FP function of degree m, written FP $m(X)$, the indices 1,...,m are replaced with the powers p_1, ..., p_m, giving:

$$\mathrm{FP}\, m(X) = \beta_1 X^{P1} + \beta_2 X^{P2} + \cdots + \beta_m X^{Pm} \tag{2}$$

As proposed by Royston and Altman (1994), the powers $p_1, \ldots, p_m$ are chosen from a restricted set, $S = \{-2, -1, -0.5, 0, 0.5, 1, 2, 3\}$, where X^0 denotes log*X*. No subsequent changes to S have been adopted. The set includes no transformation ($p = 1$) and the reciprocal, logarithmic, square root, and square transformations. FP2 functions take the form:

$$\mathrm{FP2} = \beta_1 X^{\mathrm{P1}} + \beta_2 X^{\mathrm{P2}} \tag{3}$$

Or

$$\mathrm{FP2} = \beta_1 X^{\mathrm{P1}} + \beta_2 X^{\mathrm{P1}} \log X \tag{4}$$

The latter being the so-called repeated-powers function obtained in the mathematical limit $p_2 \rightarrow p_1$ (Royston and Sauerbrei 2005).

Consider a linear regression model:

$$Y = X\beta_2 + e \tag{5}$$

where Y is an n-vector of observed responses, X is an $n \times k$ matrix representing $k$covariates, β is a k-vector of parameters and e is an n-vector of residuals with$E(e) = 0$, var$(e) = \sigma^2 I$. The ordinary least squares (OLS) estimator of β is $\hat{\beta} = (X'X)^{-1}X'Y$, and the vector of fitted values is $\hat{Y} = X\hat{\beta} = HY$.. The matrix H=X(X'X)−1X' is usually called the "hat" matrix. Its diagonal elements $H_{ij}x_i(X'X)^{-1}x_i'$ I are termed leverages. In the special case of a single covariate, x, the leverages are as follows given (Belsley et al. 1980; Royston and Sauerbrei 2007):

$$h_i = \frac{1}{n} + \frac{(x_i - \bar{x})^2}{\sum_{j=1}^{n} (x_j - \bar{x})^2} \tag{6}$$

The leverage is then a linear function of the squared distance of x_i from the covariate mean, $\bar{x}$.

Multivariable fractional polynomial modeling

The central problem of modeling is the joint effect of several covariates, at least two of which are continuous and some may be categorical or binary. At the same time, the model must be simplified, both/either dropping nonsignificant variables and/or by reducing the complexity of FP functions fitted to continuous covariates. A solution to finding a model in this situation was proposed by Royston and Altman (1994) and further refined by Sauerbrei and Royston (1999). In essence, it is impractical to try all combinations of powers for finding the best-fitting set of FP functions for all of the candidate continuous variables, and an iterative algorithm is needed. This was termed the MFP procedure by Sauerbrei and Royston (1999). It combines backward elimination of variables with a search for the best FP functions of continuous covariates. First, a significance level such as 0.05 is chosen. Next, the fitting order is determined by fitting a model linear in all variables and eliminating each variable singly, which is the first step of a conventional backward elimination procedure. The *p* values from the tests are used to order the variables, from the most to the least significant. Having imposed an order, each *X* is considered in turn. If an *X* is categorical or binary, a standard likelihood ratio test is performed to determine whether it should be dropped or not. If an *X* is continuous, the FP model selection procedure is applied to determine the best FP (or to eliminate *X*). The procedure cycles through the variables in the same order until the selected variables and FP functions do not change. Typically, two or three cycles are required for convergence (Royston and Sauerbrei 2005).

ROBUST REGRESSION

The models described in linear regression models are based on certain assumptions, such as a normal distribution of errors in the observed responses. If the distribution of errors is asymmetric or prone to outliers, model assumptions are invalidated, and parameter estimates, confidence intervals, and other computed statistics become unreliable. The Statistics Toolbox function robust fit is useful in these cases. The function implements a robust fitting method that is less sensitive than OLS to large changes in small parts of the data (Matlab Statistics Toolbox 2010).

Robust regression works by assigning a weight to each data point. Weighting is done automatically and iteratively using a process called *iteratively reweighted least squares*. In the first iteration, each point is assigned equal weight and model coefficients are estimated using ordinary least squares. At subsequent iterations, the weights are recomputed so that the points farther from model predictions in the previous iteration are given a lower weight. The model coefficients are then recomputed using weighted least squares (WLS). The process continues until the values of the coefficient estimates converge within a specified tolerance (Matlab Statistics Toolbox 2010).

Robust demo demonstrates the difference between ordinary (least squares) regression and robust regression. It displays a scatter plot of *X* and *Y* values, where*Y* is roughly a linear unction of *X*, but one point is an outlier (it falls far from the line). The bottom of the figure shows the fitted equations using least squares and robust fitting, plus an estimate of the root mean squared error from both. Leverage is a measure of how much influence that point has on the least squares fit. Weight is the weight that point was given in the robust fit. Robust demo (*X*, *Y*) uses *X* and Y data that supply, in place of the sample data supplied with the function (Matlab Statistics Toolbox 2010).

Bayesian regression

Bayesian regression, the theory specialized adaptation is including the development of multivariate regression models, which clearly consider two sources of previous and experimental information.

Bayesian regression analysis is including development of model or prediction of the relations among variables. For example, a prediction equation can be the following linear form (Eq. 7).

$$y = b_0 + b_1x_1 + b_2x_2 + \cdots + b_px_p \tag{7}$$

where *Y* is the dependent variable, independent variables are predicted by X_i, and the aim is numerical determine of b_j coefficients.

The Bayesian model makes possible the decision in short time with development of data and judgments and continue of modeling.

Suppose that we are interested in estimating θ from data $y = \{y_1, \ldots, y_n\}$ using a statistical model described by a density $p(y|\theta)$. Bayesian philosophy states that θcannot be determined exactly, and uncertainty about the parameter is expressed through probability statements and distributions. Also, we can say that θ follows a normal distribution with mean 0 and variance 1, if it is believed that this distribution best describes the uncertainty associated with the parameter. The following steps describe the essential elements of Bayesian inference:

1. A probability distribution for θ is formulated as $\pi(\theta)$, which is known as the *prior* distribution, or just the prior. The prior distribution expresses beliefs (e.g., on the mean, the spread, and the skewness) about the parameter before examining the data.

2. Given the observed data y, choose a statistical model $p(y|\theta)$ to describe the distribution of y given θ.

3. Update of beliefs about θ by combining information from the prior distribution and the data through the calculation of the*posterior* distribution, $p(y|\theta)$.

The third step is carried out using Bayes' theorem, which enables to combine the prior distribution and the model in the following way:

$$p(y|\theta) = \frac{p(\theta.y)}{p(y)} = \frac{p(y|\theta)\pi(\theta)}{p(y)} = \frac{p(y|\theta)\pi(\theta)}{\int p(y|\theta)\pi(\theta)\mathrm{d}\theta} \quad (8)$$

The quantity

$$p(y) = \int p(y|\theta)\pi(\theta)\mathrm{d}\theta \quad (9)$$

is the normalizing constant of the posterior distribution. This quantity *p(y)* is also the marginal distribution of *y*, and it is sometimes called the marginal distribution of the data. The likelihood function of θ is any function proportional to $p(y|\theta)$; that is $L(\theta) \propto p(y|\theta)$.

In this study, daily grass reference ET (ET_0) is determined using the FAO Penman–Monteith (FAO-PM) approach (Allen et al. 1998). There are many studies that applied FAO-PM model as a base model for calculation of reference ET in different regions of Iran (Rahimi et al. 2014; Valipour 2012c, d, e, f, 2015d; Valipour and Eslamian 2014). The FAO-PM equation for a grass reference crop is defined as:

$$\mathrm{ET}_0 = \frac{0.408\Delta(R_n - G) + \gamma(900/(T + 273))u_2(e_s - e_a)}{\Delta + \gamma(1 + 0.34u_2)} \quad (10)$$

where ET_0 is the evapotranspiration (mm day^{-1}), R_n is the net radiation (MJ m^2 day^{-1}), G is the soil heat flux (MJ m^2 day^{-1}), T is the mean daily air temperature (°C), u_2 is the mean daily wind speed at 2 m height (m s^{-1}), $e_s - e_a$ is the saturation vapor pressure deficit (kPa °C^{-1}), and γ is the psychometric constant (kPa °C^{-1}).

In this study, three models were performed including MFP, Bayesian regression, and robust regression models. All of the models were designed with three parameters as input data, including solar radiation (Ra), mean temperature (T_{mean}), and $T_{max} - T_{min}$.

STUDY AREA

In this study, three synoptic stations have been used that have at least 14-year data to avoid the errors due to the problem of short record length. These stations are located on central part of Iran with an arid environment. Figure 1 and Table 1 show the position and geographic characteristics of

Esfahan, Kashan, and Ardestan stations, respectively. All of the information is in monthly scale and has been collected from Islamic Republic of Iran Meteorological Organization (IRIMO) at this link: http://irimo.ir/eng/index.php.

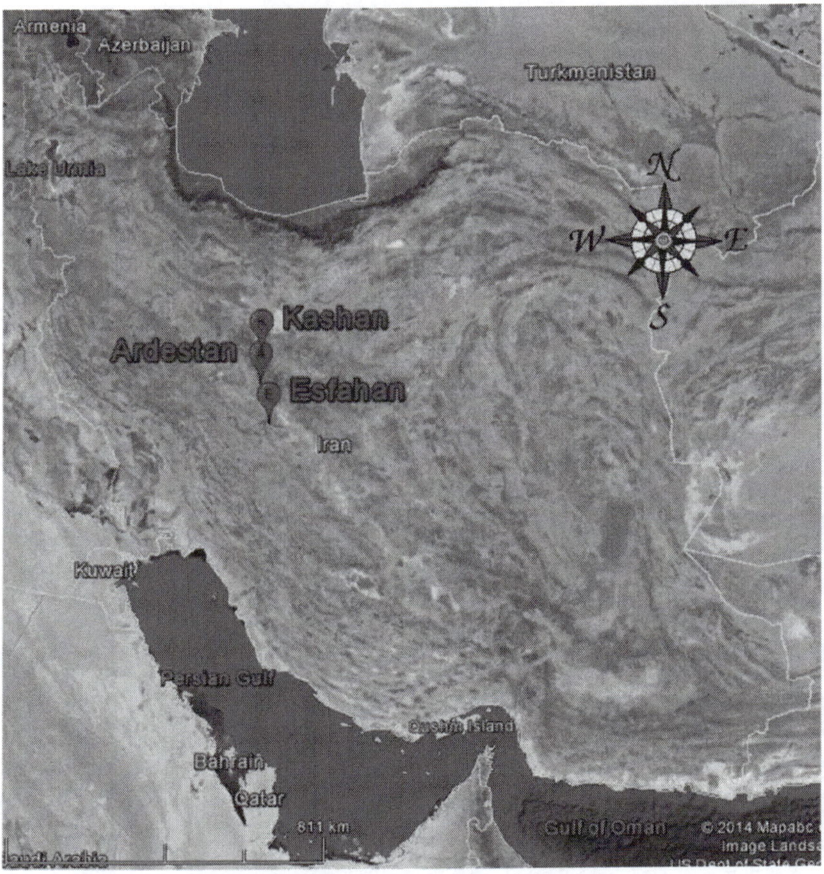

Figure 1. Position of Esfahan, Kashan, and Ardestan in Iran

Table 1. Geographic characteristics of the stations.

Station	Latitude	Longitude	Elevation m.a.s.l	Record duration	Annual precipitation (mm)	Climate (de Martonne)
Esfahan	32°37′N	51°40′E	1550.4	1961–2005	122.8	Arid
Kashan	33°59′N	51° 27′E	982.3	1966–2005	138.4	Arid
Ardestan	33°23′N	52° 23′E	1252.4	1992–2005	115.8	Arid

To evaluate the performance of these models in monthly ET_0 estimates, between the calculated and predicted ET, the root mean square error (RMSE) is applied as given:

$$\mathrm{RMSE} = \sqrt{\frac{1}{n}\sum_{i=1}^{n}\left(\left(\frac{Y_{\mathrm{obs}} - Y_{\mathrm{calc}}}{Y_{\mathrm{obs}}}\right)^2\right.} \quad (11)$$

where Y_{calc} is the calculated value and Y_{est} is estimation of the models.

Table 2. Statistical parameters of the MFP model for ET prediction

Stations	R^2	RMSE
Esfahan	0.96	9.48
Kashan	0.98	7.58
Ardestan	0.98	8.74

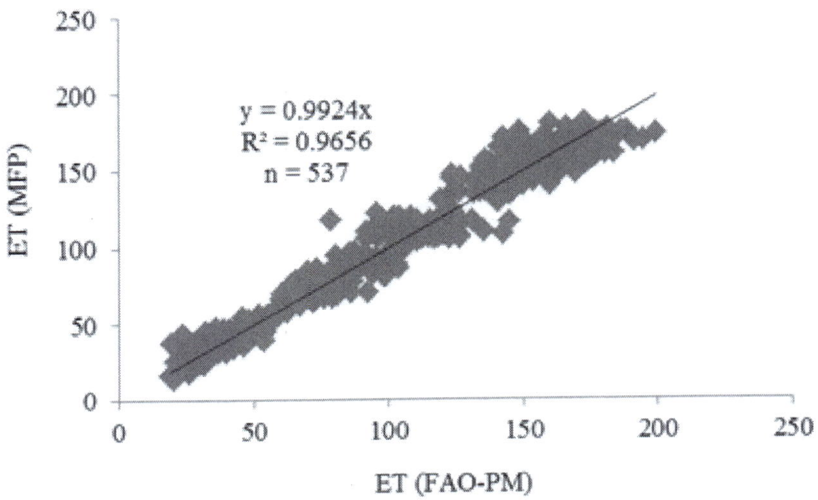

Figure 2. Comparison of the ET predicted with MFP model versus the calculated values by FAO-PM for Esfahan station

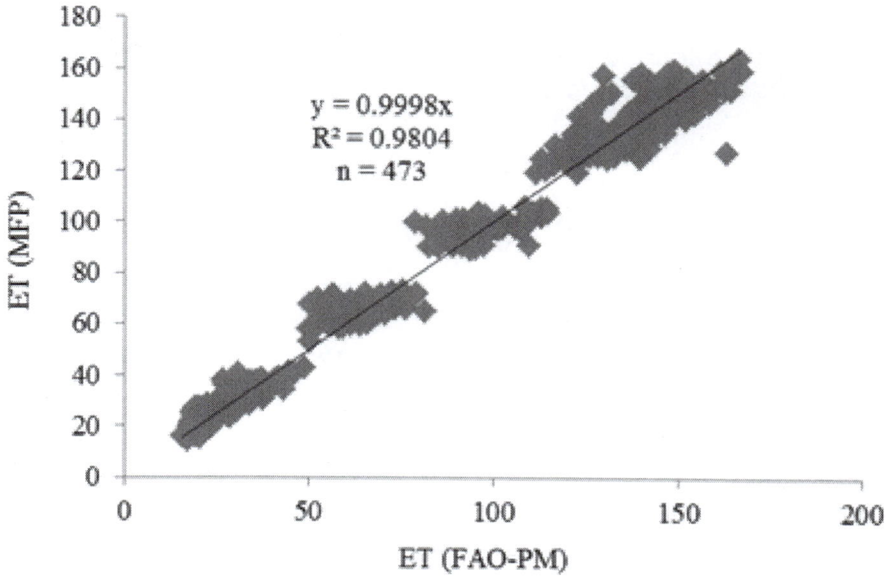

Figure 3. Comparison of the ET predicted with MFP model versus the calculated values by FAO-PM for Kashan station

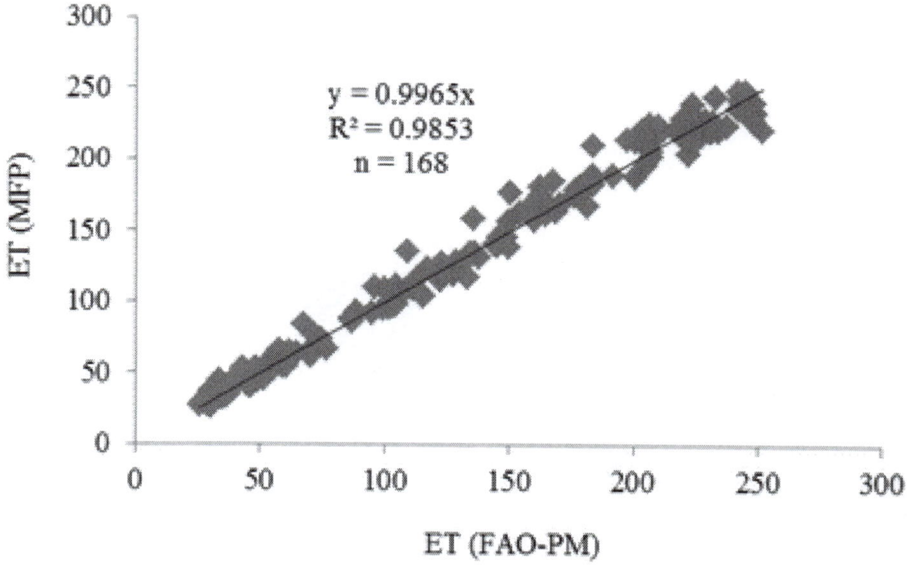

Figure 4. Comparison of the ET predicted with MFP model versus the calculated values by FAO-PM for Ardestan station

RESULTS AND DISCUSSIONS

Multivariable fractional polynomial modeling

The results of statistical parameters of MFP model are given in Table 2. In addition, Figs. 2, 3 and 4 show the predicted values of ET versus the calculated values of ET.

To evaluate the MFP model, the amount of ET estimated by this model was plotted versus the calculated values. These results are shown for Esfahan, Kashan, and Ardestan stations in Figs. 2, 3 and 4, respectively. The best performance of the MFP model was obtained for Kashan than Esfahan and Ardestan stations.

Considering the obtained results, in this study, it could be stated that use of MFP model can predict ($R^2 > 0.95$) the amount of ET in different stations, and this application can be added to the other applications of MFP model.

The formula obtained from MFP model for Esfahan, Kashan, and Ardestan stations are shown in Eqs. 12, 13, and 14, respectively.

$$\begin{aligned} \mathrm{ET} = & -323.91 + 54.6\left(\frac{\mathrm{Ra}}{1000}\right)^{-1} + 293.42\left(\frac{\mathrm{Ra}}{1000}\right)^{0.5} \\ & + 23.71\left(\frac{T_{\mathrm{mean}} + 3.4}{10}\right) + 11.74\left(\frac{T_{\max} - T_{\min}}{10}\right)^{3} \\ & - 16.62\left(\frac{T_{\max} - T_{\min}}{10}\right)^{3} \mathrm{Ln}\left(\frac{T_{\max} - T_{\min}}{10}\right) \end{aligned} \quad (12)$$

$$\begin{aligned} \mathrm{ET} = & -13.304 + 63.466\left(\frac{\mathrm{Ra}}{1000}\right)^{2} + 4.172\left(\frac{T_{\mathrm{mean}}}{10}\right)^{2} \\ & + 14.324\left(\frac{T_{\max} - T_{\min}}{10}\right) \end{aligned} \quad (13)$$

$$\begin{aligned} \mathrm{ET} = & -39.147 + 62.267\left(\frac{\mathrm{Ra}}{1000}\right)^{2} + 9.524\left(\frac{T_{\mathrm{mean}}}{10}\right)^{2} \\ & + 56.422\left(\frac{T_{\max} - T_{\min}}{10}\right) \end{aligned} \quad (14)$$

ROBUST REGRESSION

The results of statistical parameters of robust regression model are given in Table 3. In addition, Figs. 5, 6 and 7 show the predicted values of ET versus the calculated values of ET.

To evaluate the robust regression model, the amount of ET estimated by this model was plotted versus the calculated values. These results are shown for Esfahan, Kashan, and Ardestan stations in Figs. 5, 6 and 7, respectively. The maximum value of the correlation coefficient was obtained for Ardestan; however, the minimum value of RMSE belongs to Kashan.

Table 3. Statistical parameters of the robust regression model for ET prediction

Stations	R^2	RMSE
Esfahan	0.95	10.59
Kashan	0.96	8.24
Ardestan	0.97	12.07

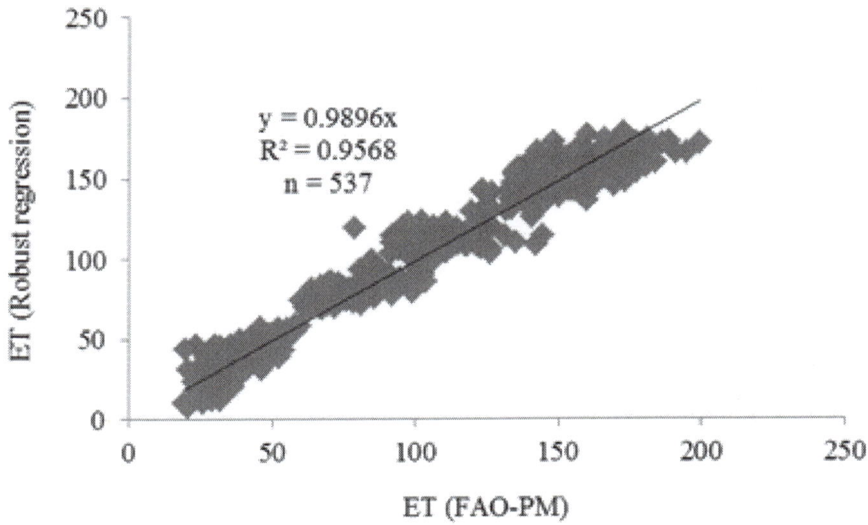

Figure 5. Comparison of the ET predicted with robust regression model versus the calculated values by FAO-PM for Esfahan station

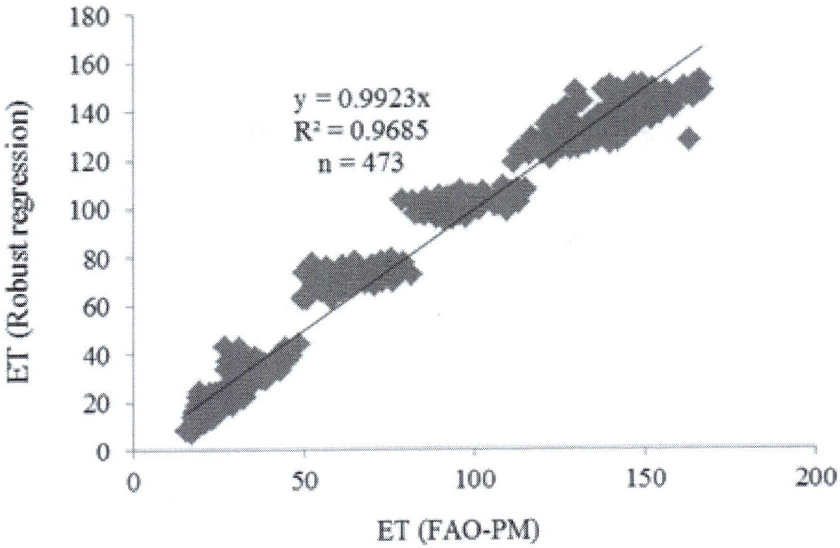

Figure 6. Comparison of the ET predicted with robust regression model versus the calculated values by FAO-PM for Kashan station

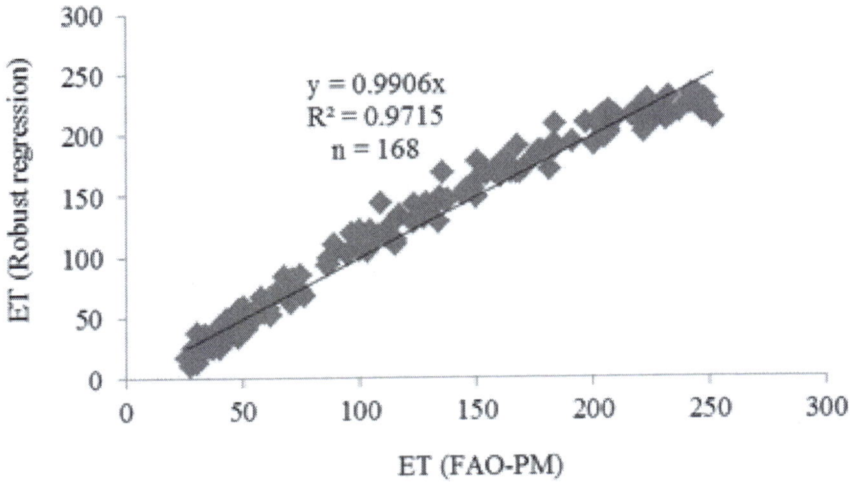

Figure 7. Comparison of the ET predicted with robust regression model versus the calculated values by FAO-PM for Ardestan station

The formula obtained from robust regression model for Esfahan, Kashan, and Ardestan stations are shown in Eqs. 15, 16, and 17, respectively.

$$ET = -19.6047 + 2.8297T_{\text{mean}} - 0.7866(T_{\text{max}} - T_{\text{min}}) + 0.0753\text{Ra} \quad (15)$$

$$\text{ET} = -51.9819 + 2.022T_{\text{mean}} - 0.0465(T_{\text{max}} - T_{\text{min}}) + 0.1061\text{Ra} \quad (16)$$

$$\text{ET} = -87.4207 + 3.9657T_{\text{mean}} + 2.381(T_{\text{max}} - T_{\text{min}}) + 0.1195\text{Ra} \quad (17)$$

Bayesian regression

The results of statistical parameters of Bayesian regression model are given in Table 4. In addition, Figs. 8, 9, and 10 show the predicted values of ET versus the calculated values of ET.

To evaluate the Bayesian regression model, the amount of ET estimated by this model was plotted versus the calculated values. These results are shown for Esfahan, Kashan, and Ardestan stations in Figs. 8, 9, and 10, respectively. Similar to the robust regression analysis, the maximum value of the correlation coefficient was obtained for Ardestan; however, the minimum value of RMSE belongs to Kashan. It should be noted that performance of the Bayesian model for Kashan (Fig. 9) is the only overestimating observed in all models among all stations.

Table 4. Statistical parameters of the Bayesian regression model for ET prediction

Station	R^2	RMSE
Esfahan	0.95	10.59
Kashan	0.96	8.84
Ardestan	0.97	12.05

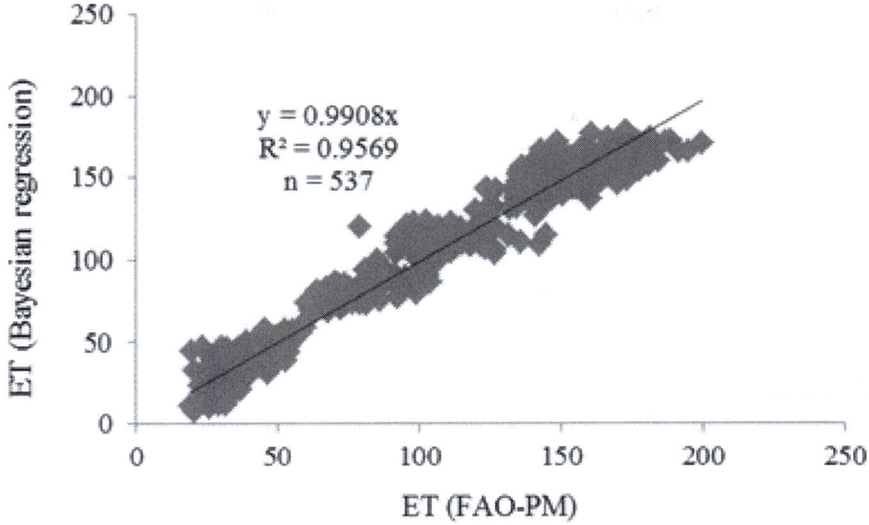

Figure 8. Comparison of the ET predicted with Bayesian regression model versus the calculated values by FAO-PM for Esfahan station

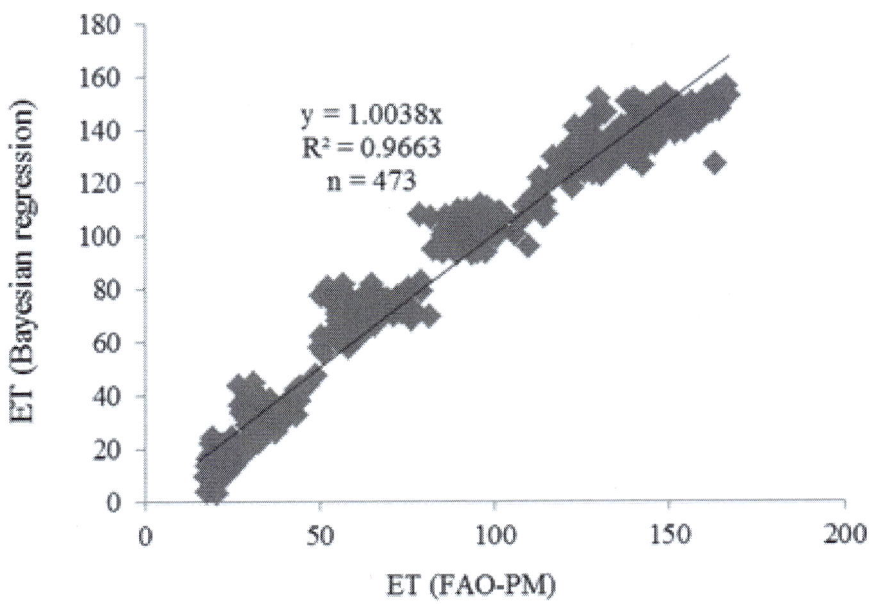

Figure 9. Comparison of the ET predicted with Bayesian regression model versus the calculated values by FAO-PM for Kashan station

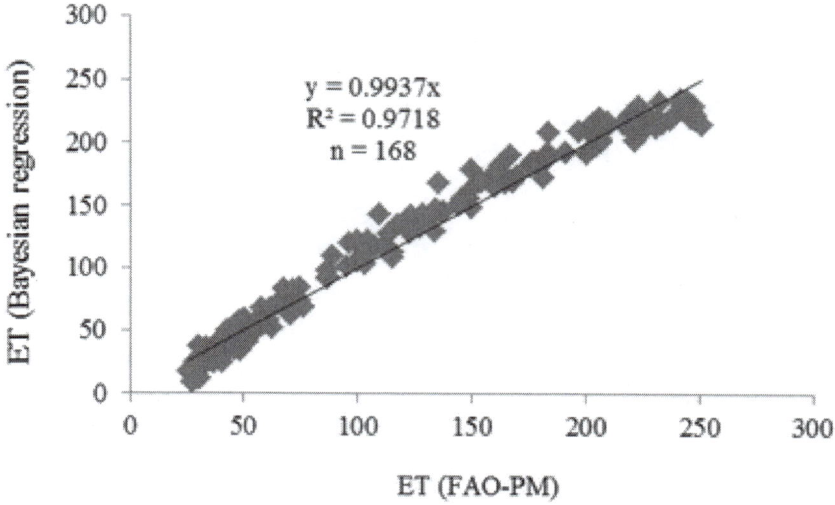

Figure 10. Comparison of the ET predicted with Bayesian regression model versus the calculated values by FAO-PM for Ardestan station

The formula obtained from Bayesian regression model for Esfahan, Kashan, and Ardestan stations are shown in Eqs. 18, 19, and 20, respectively.

$$\mathrm{ET} = -20.2773 + 0.0769\mathrm{Ra} - 0.7886(T_{\max} - T_{\min}) + 2.7749T_{\text{mean}} \tag{18}$$

$$\mathrm{ET} = -61.0469 + 0.0961\mathrm{Ra} + 0.9314(T_{\max} - T_{\min}) + 2.2568T_{\text{mean}} \tag{19}$$

$$\mathrm{ET} = -87.9839 + 0.1166\mathrm{Ra} + 2.4713(T_{\max} - T_{\min}) + 4.0873T_{\text{mean}} \tag{20}$$

According to Tables 2, 3, and 4, the best results obtained from the MFP model with the highest regression coefficient and the lowest RMSE. The MFP model is more flexible than the ordinary polynomial methods and more resistant than nonparametric regression methods.

Figures 11 and 12 are plotted to better comparison of the performance of the models used with respect to the average values of the evaluation indices.

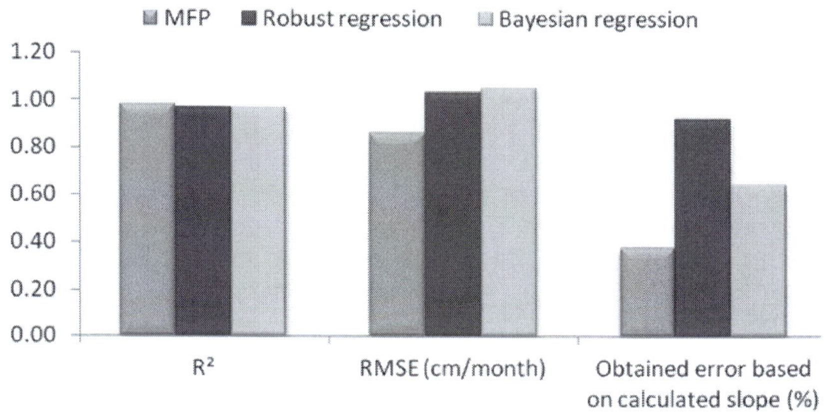

Figure 11. Comparison of the models used according to the average of the evaluation indices

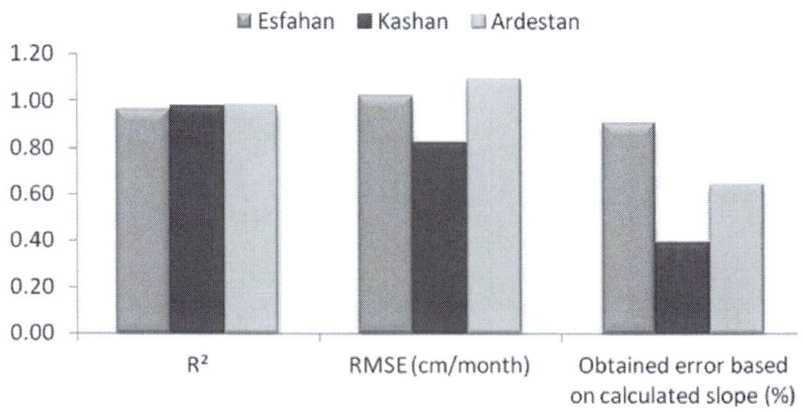

Figure 12. Comparison of the average of evaluation indices for area studies

According to Fig. 11, once again the MFP model is introduced as the best model for estimation of reference ET in the study areas due to the highest correlation coefficient and the lowest RMSE and slope error. It should be noted that the slope error was obtained based on the coefficient of the suggested equations in Figs. 2, 3, 4, 5, 6, 7, 8,9, and 10.

According to Fig. 12, the models used estimate reference ET in Kashan better than Esfahan and Ardestan. Table 1 confirms this result because the annual precipitation of Kashan (138.4 mm) is more than both Esfahan (122.8 mm) and Ardestan (115.8 mm). In addition, compared with Esfahan (1550.4 masl) and Ardestan (1252.4 masl), the lowest elevation belongs to Kashan (982.3 masl). Therefore, application of the models is proposed for all regions in the world with weather conditions close to Kashan, and this leads to the best performance as well as reduction of uncertainty for estimation of reference ET.

Compared with the previous works (in these three regions), the obtained results underline better performance of MFP, robust regression, and Bayesian regression in Esfahan, Kashan, and Ardestan. Azizi et al. (2010) applied multivariations regression model for estimating reference ET in Esfahan province. The authors used four weather parameters (temperature, air vapor pressure, relative humidity, and wind speed) to estimate reference ET and claimed that use of this approach led to correlation coefficients equal to 0.998 and 0.992, in Esfahan and Kashan, respectively. In the other study, Heydari et al. (2013) evaluated different methods include Penman FAO, Blaney–Criddle, Linacre, Thorenthwaite, Hargreaves–Samani, Irmak, Turc, and Jensen–Haise in Ardestan, and then their results were compared with FAO-PM method. The results indicated that the Blaney–Criddle model is the most appropriate for this particular study area, with an average RMSE of 1.34 mm day^{-1} and MAE of 1.00 mm day^{-1} and a correlation coefficient of 0.952. The obtained results can be expanded for other arid regions with similar wheatear conditions.

SUMMARY AND CONCLUSIONS

In this research, the capacity of different models for evapotranspiration prediction was investigated. Evapotranspiration is a complex and nonlinear phenomenon because it depends on several interacting climatological factors, such as temperature, humidity, wind speed, radiation, type, and growth stage of the crop (Jain et al. 2008; Ojha and Bhakar 2012; Valipour 2012a, b, 2013c, d, e, f, g, h, Valipour 2014b). The MFP model is an effective tool to model nonlinear systems. The obtained results of predictions are in good agreement with the calculated ET at all stations. The results showed that the accuracy of MFP model was greater than the other two models in estimation of the amount of ET. Namely, the MFP model has more regression coefficient values and the less prediction errors than the other two models. Therefore, the suggestion is that these models, especially the MFP, could apply for other climatic and hydrologic modeling.

One of the most important results of the current investigation is estimation of reference ET by using only two weather variables including temperature and radiation, with an acceptable accuracy. Therefore, the formulae obtained are preferred to (1) other models in which reference ET was estimated using only these to variables with a future error as well as (2) other models in which reference ET was estimated using more weather variables such as relative humidity, wind speed, elevation, and precipitation.

REFERENCES

1. Abtew W (1996) Evapotranspiration measurements and modeling for three wetland systems in South Florida. J Water Resour Bull 32(3):465–473
2. Abtew W, Obeysekera J (1995) Lysimeter study of evaporation of cattails and comparison of 3 estimation methods. Trans ASAE 38(1):121–129
3. Allen RG (1986) A Penman for all seasons. J Irrig Drain Eng 112(4):348–368
4. Allen RG, Jensen ME, Wright JL, Burman RD (1989) Operational estimates of reference evapotranspiration. J Agron 81:650–662
5. Allen RG, Pereira LS, Raes D, Smith M (1998) Crop evapotranspiration guidelines for computing crop water requirements. FAO irrigation and drainage. Paper no. 56, FAO, Rome
6. Azizi G, Hanafi A, Soltani M (2010) Evaluation of potential evapotranspiration via multi-variations regression in Esfahan province. Geography 11(3):77–92 **(in Persian)**
7. Bachour R, Walker WR, Ticlavilca AM, McKee M, Maslova I (2014) Estimation of spatially distributed evapotranspiration using remote sensing and a relevance vector machine. J Irrig Drain Eng 140(8):04014029
8. Banihabib ME, Valipour M, Behbahani SMR (2012) Comparison of autoregressive static and artificial dynamic neural network for the forecasting of monthly inflow of Dez reservoir. J Environ Sci Technol 13(4):1–14
9. Belsley DA, Kuh E, Welsch RE (1980) Regression diagnostics. Wiley, New York
10. Block P, Rajagopalan B (2009) Statistical-dynamical approach for streamflow modeling at malakal, sudan, on the white Nile river. J Hydrol Eng 14(2):185–196
11. Box GEP, Tidwell PW (1962) Transformation of the independent variables. Technometrics 4:531–550

12. Chiew FHS, Kamaladassa NN, Malano HM, McMahon TA (1995) Penman-Monteith, FAO-24 reference crop evapotranspiration and class-A pan data in Australia. J Agr Water Manage 28:9–21

13. Chulani S, Boehm B, Steece B (1998) Calibrating software cost models using bayesian analysis. University of Southern California. Henry Salvatori, Los Angeles, CA 90089-0781. USC-CSE

14. Coleman D, Holland P, Kaden N, Klema V (1980) A system of subroutines for iteratively reweighted least squares computations. ACM Trans Math Softw 6:327–336

15. De Souza F, Yoder RE (1994) ET estimation in the northeast of Brazil: Hargreaves or penman-monteith equation? In: Proceedings of technical paper ASAE international winter meeting, American Society of Agricultural Engineers, St. Joseph, Mich

16. Demoster AP, Laird NM, Rubin DB (1980) Iteratively reweighted least squares for linear regression when errors are normal/independent distributed. Multivariate analysis V. Krishnaiah PR, North-Holland, New York, pp 35–37

17. Douglas EM, Jacobs JM, Sumner DM, Ray RL (2009) A comparison of models for estimating potential evapotranspiration for Florida land cover types. J Hydrol 373:366–376

18. Ershadi A, McCabe MF, Evans JP, Mariethoz G, Kavetski D (2013) A Bayesian analysis of sensible heat flux estimation: quantifying uncertainty in meteorological forcing to improve model prediction. Water Resour Res 49:2343–2358

19. Eslamian SS, Abedi-Koupai J, Amiri MJ, Gohari SA (2009) Estimation of daily reference evapotranspiration using support vector machines and artificial neural networks in Greenhouse. Res J Environ Sci 3:439–447

20. Ghosh A, Joshi PK (2014) Hyperspectral imagery for disaggregation of land surface temperature with selected regression algorithms over different land use land cover scenes. ISPRS J Photogramm Remot Sens 96:76–93

21. Heydari MM, Abbasi A, Heydari M (2013) Estimation of evapotranspiration in Ardestan, center of Iran. World Appl Sci J 21(2):230–236

22. Iizumi T, Nishimori M, Yokozawa M, Kotera A, Duy Khang N (2012) Statistical downscaling with Bayesian inference: estimating global solar radiation from reanalysis and limited observed data. Int J Climatol 32:464–480

23. Izadifar Z, Elshorbagy A (2010) Prediction of hourly actual evapotranspiration using neural networks, genetic programming, and statistical models. Hydrol Process 24:3413–3425

24. Jain SK, Nayak PC, Sudheer KP (2008) Models for estimating evapotranspiration using artificial neural networks, and their physical interpretation. Hydrol Process 22:2225–2234

25. Jensen ME, Burman RD, Allen RG (1990) Evapotranspiration and irrigation water requirements. ASCE manual and report on engineering practice no. 70. ASCE, New York

26. Jones JW, Keating BA, Porter CH (2001) Approaches to modular model development. Agr Syst 70:421–443

27. Joshi D, St-Hilaire A, Daigle A, Ouarda TB (2013) Databased comparison of sparse bayesian learning and multiple linear regression for statistical downscaling of low flow indices. J Hydrol 488:136–149

28. Kumar M, Raghuwanshi NS, Singh R, Wallender WW, Pruitt WO (2002) Estimating evapotranspiration using artificial neural network. J Irrig Drain Eng 128(4):224–233CrossRef

29. Liang Z, Wang D, Guo Y, Zhang Y, Dai R (2013) Application of bayesian model averaging approach to multimodel ensemble hydrologic forecasting. J Hydrol Eng 18(11):1426–1436

30. Mahdizadeh Khasraghi M, Gholami Sefidkouhi MA, Valipour M (2015) Simulation of open- and closed-end border irrigation systems using SIRMOD. Arch Agron Soil Sci 61(7):929–941

31. Marshall L, Nott D, Sharma A (2007) Towards dynamic catchment modelling: a Bayesian hierarchical mixtures of experts framework. Hydrol Process 21:847–861

32. Matlab Statistics Toolbox (2010) User manual MATLAB STATISTICS TOOLBOX 7. http://somemanuals.com/MATLAB,STATISTICS-TOOLBOX-7,ICDFDDF.PDF

33. McNaughton KG, Jarvis PG (1984) Using Penman-Monteith equation predictively. Agr Water Manage 8:263–278

34. Ojha S, Bhakar SR (2012) Estimation of evapotranspiration for wheat crop using artificial neural network. J Indian Water Resour Soci 32:13–21

35. Pejic B, Aksic M, Mackic K, Sekularac G (2015) Response of potato to water stress in Southern Serbia. Austin J Irrig 1(1):1001

36. Rahimi S, Gholami Sefidkouhi MA, Raeini-Sarjaz M, Valipour M (2014) Estimation of actual evapotranspiration by using MODIS

images (a case study: tajan catchment). Arch Agron Soil Sci 61(5):695–709

37. Rahimikhoob A (2009) Estimating daily pan evaporation using artificial neural network in a semi-arid environment. Theor Appl Climatol 98:101–105

38. Rathnayake RC (2010) Bayesian inference and computation. Research papers. Paper 29. http://opensiuc.lib.siu.edu/gs_rp/29

39. Rosenberry DO, Winter TC, Buso DC, Likens GE (2007) Comparison of 15 evaporation methods applied to a small mountain lake in the Northeastern USA. J Hydrol 340:149–166

40. Royston P, Altman DG (1994) Regression using fractional polynomials of continuous covariates: parsimonious parametric modelling (with discussion). Appl Stat 43:429–467

41. Royston P, Sauerbrei W (2005) Building multivariable regression models with continuous covariates in clinical epidemiology with an emphasis on fractional polynomials. Meth Inf Med 44:561–571

42. Royston P, Sauerbrei W (2007) Improving the robustness of fractional polynomial models by preliminary covariate transformation: a pragmatic approach. Comput Stat Data Anal 51:4240–4253

43. Sauerbrei W, Royston P (1999) Building multivariable prognostic and diagnostic models: transformation of the predictors using fractional polynomials. J Roy Stat Soc (Series A) 162:71–94 Corrigendum: J Roy Stat Soc (2002) (Series A) 165:399–400

44. Schuch UK, Burger DW (1997) Water use and crop coefficients of woody ornamentals in containers. J Am Soc Hort Sci 122:727–734

45. Singh K (2014) A review on software cost estimation models and techniques. IJETI Int J Eng Technol Innovat 1(3):17–26

46. Tabari H, Kisi O, Ezani A, Hosseinzadeh Talaee P (2012) SVM, ANFIS, regression and climate based models for reference evapotranspiration modeling using limited climatic data in a semi-arid highland environment. J Hydrol 444:78–89

47. Ticlavilca AM, McKee M, Walker WR (2013) Real-time forecasting of short-term irrigation canal demands using a robust multivariate Bayesian learning model. Irrig Sci 31(2):151–167

48. Tzimopoulos C, Mpallas L, Papaevangelou G (2007) Estimation of evapotranspiration using fuzzy rules and comparison with the Blaney-Criddle method. In: Water Resour Manag: new approaches and technologies. Edited by: European Water Resources Association, pp 245–252 Chania, Crete, Greece, June 24–28

49. Valipour M (2012a) Hydro-module determination for Vanaei Village in Eslam Abad Gharb. Iran. ARPN J Agr Biol Sci 7(12):968–976

50. Valipour M (2012b) Ability of box-Jenkins models to estimate of reference potential evapotranspiration (A case study: Mehrabad synoptic station, Tehran, Iran). IOSR J Agr Veter Sci 1(5):1–11

51. Valipour M (2012c) Comparison of Surface Irrigation Simulation Models: full Hydrodynamic, Zero Inertia. Kinematic Wave. J Agr Sci 4(12):68–74

52. Valipour M (2012d) Scrutiny of pressure loss, friction slope, inflow velocity, velocity head, and reynolds number in Center Pivot. Int J Adv Sci Technic Res 2(5):703–711

53. Valipour M (2012e) Determining possible optimal values of required flow, nozzle diameter, and wetted area for linear traveling laterals. Int J Eng Sci 1(1):37–43

54. Valipour M (2012f) Sprinkle and trickle irrigation system design using tapered pipes for pressure loss adjusting. J Agr Sci 4(12):125–133

55. Valipour M (2013a) Use of surface water supply index to assessing of water resources management in colorado and oregon, US. Adv Agr Sci Eng Res 3(2):631–640

56. Valipour M (2013b) Estimation of surface water supply index using snow water equivalent. Adv Agr Sci Eng Res 3(1):587–602

57. Valipour M (2013c) Necessity of irrigated and rainfed agriculture in the world. Irrig Drain Sys Eng S9:e001. http://www.omicsgroup.org/journals/2168-9768/pdfdownload.php?download=2168-9768-S9-e001.pdf&&aid=12800

58. Valipour M (2013d) Increasing irrigation efficiency by management strategies: cutback and surge irrigation. ARPN J Agr Biol Sci 8(1):35–43

59. Valipour M (2013e) Evolution of irrigation-equipped areas as share of cultivated areas. Irrig Drain Sys Eng 2(1):e114. http://www.omicsgroup.org/journals/evolution-of-irrigationequipped-areas-as-share-of-cultivated-areas-2168-9768.1000e114.php?aid=12797

60. Valipour M (2013f) Need to update of irrigation and water resources information according to the progresses of agricultural knowledge. Irrig Drain Sys Eng S10:e001. http://www.omicsgroup.org/journals/need-to-update-of-irrigation-and-water-resources-information-according-to-the-progresses-of-agricultural-knowledge-2168-9881.S10-e001.php?aid=15774

61. Valipour M (2013g) Comparison of different drainage systems usable for solution of environmental crises in soil. In: The 1st international conference on environmental crises and its solutions, Kish Island, Iran

62. Valipour M (2013h) Scrutiny of inflow to the drains applicable for improvement of soil environmental conditions. In: The 1st international conference on environmental crises and its solutions, Kish Island, Iran

63. Valipour M (2014a) Analysis of potential evapotranspiration using limited weather data. Appl Water Sci. Assessed 27 September 2014.

64. Valipour M (2014b) Application of new mass transfer formulae for computation of evapotranspiration. J Appl Water Eng Res 2(1):33–46

65. Valipour M (2014c) Variations of irrigated agriculture indicators in different continents from 1962 to 2011. Adv Water Sci Technol 1(1):1–14

66. Valipour M (2014d) Assessment of different equations to estimate potential evapotranspiration versus FAO Penman Monteith method. Acta Adv Agr Sci 2(11):14–27

67. Valipour M (2014e) Irrigation status of Americas. Adv Appl Agr Sci 2(12):56–72

68. Valipour M (2014f) Handbook of irrigation engineering problems. OMICS Group eBooks, Foster City. http://www.esciencecentral.org/ebooks/handbook-of-irrigation-engineering-problems/pdf/handbook-of-irrigation-engineering-problems.pdf

69. Valipour M (2014g) Handbook of drainage engineering problems. OMICS Group eBooks, Foster City. http://www.esciencecentral.org/ebooks/handbook-of-drainage-engineering-problems/pdf/handbook-of-drainage-engineering-problems.pdf

70. Valipour M (2014h) Handbook of hydraulic engineering problems. OMICS Group eBooks, Foster City. http://www.esciencecentral.org/ebooks/handbook-of-hydraulic-engineering-problems/pdf/handbook-of-hydraulic-engineering-problems.pdf

71. Valipour M (2014i) Handbook of hydrologic engineering problems. OMICS Group eBooks, Foster City. http://www.esciencecentral.org/ebooks/handbook-of-hydrologic-engineering-problems/pdf/handbook-of-hydrologic-engineering-problems.pdf

72. Valipour M (2014j) Handbook of water engineering problems. OMICS Group eBooks, Foster City. http://www.esciencecentral.org/ebooks/handbook-of-water-engineering-problems/pdf/handbook-of-water-engineering-problems.pdf

73. Valipour M (2014k) Prediction of irrigated agriculture in Asia Pacific using FAO indices. Acta Adv Agr Sci 2(9):40–53

74. Valipour M (2014l) Pressure on renewable water resources by irrigation to 2060. Acta Adv Agr Sci 2(8):32–42

75. Valipour M (2014m) Future of agricultural water management in Europe based on socioeconomic indices. Acta Adv Agr Sci 2(7):1–18

76. Valipour M (2015a) What is the tendency to cultivate plants for designing cropping intensity in irrigated area? Adv Water Sci Technol 2(1):1–12

77. Valipour M (2015b) Assessment of important factors for water resources management in European agriculture. J Water Res Hydraul Eng 4(2):169–178

78. Valipour M (2015c) Handbook of environmental engineering problems. OMICS Group eBooks, Foster City. http://www.esciencecentral.org/ebooks/handbook-of-environmental-engineering-problems/pdf/handbook-of-environmental-engineering-problems.pdf

79. Valipour M (2015d) Calibration of mass transfer-based models to predict reference crop evapotranspiration. Appl Water Sci. Assessed 5 March 2015.

80. Valipour M (2015e) Optimization of neural networks for precipitation analysis in a humid region to detect drought and wet year alarms. Met Appl. Assessed 26 November 2015.

81. Valipour M, Eslamian S (2014) Analysis of potential evapotranspiration using 11 modified temperature-based models. Int J Hydrol Sci Technol 4(3):192–207

82. Valipour M, Mousavi SM, Valipour R, Rezaei E (2012) SHCP: soil heat calculator program. IOSR J Appl Phys 2(3):44–50

83. Valipour M, Gholami Sefidkouhi MA, Eslamian S (2015) Surface irrigation simulation models: a review. Int J Hydrol Sci Technol 5(1):51–70

84. Verstraeten WW, Veroustraete F, Feyen J (2008) Assessment of evapotranspiration and soil moisture content across different scales of observations. Sensor 8:70–117

85. Vigiak O, Bende-Mich U (2013) Estimating bootstrap and Bayesian prediction intervals for constituent load rating curves. Water Resour Res 49:8565–8578

86. Walter IA, Allen RG, Elliot R, Itenfisu D, Brown P (2002) The ASCE standardized reference evapotranspiration equation. In:

Proceeding of the Rep. Task Com. on standardized reference evapotranspiration, July 9, American Society of Civil Engineering, Reston, VA, USA, pp 1–70

87. Weib M, Menzel L (2008) A global comparison of four potential evapotranspiration equations and their relevance to stream flow modelling in semi-arid environments. Adv Geosci 18:15–23

88. Wellen C, Arhonditsis GB, Labencki T, Boyd D (2012) A Bayesian methodological framework for accommodating interannual variability of nutrient loading with the SPARROW model. Water Resour Res 48(10):W10505

89. Wrachien DD, Mambretti S (2015) Irrigation and drainage systems in flood-prone areas: the role of mathematical models. Austin J Irrig 1(1):1002

90. Wu IP (1997) A Simple Evapotranspiration Model for Hawaii: The Hargreaves Model. Department of Biosystems Engineering. Cooperative Extension Service. CTAHR Fact Sheet Engeineer's Notebook no. 106

91. Yannopoulos SI, Lyberatos G, Theodossiou N, Li W, Valipour M, Tamburrino A, Angelakis AN (2015) Evolution of water lifting devices (pumps) over the centuries worldwide. Water 7(9):5031–5060

92. Yao Y, Liang S, Li X, Hong Y, Fisher JB, Zhang N, Jiang B, Sun L, Jia K, Wang K, Chen Y, Mu Q, Feng F (2014) Bayesian multimodel estimation of global terrestrial latent heat flux from eddy covariance, meteorological, and satellite observations. J Geophys Res Atmos 119:4521–4545

93. Zhu G, Su Y, Li X, Zhang K, Li C (2013) Estimating actual evapotranspiration from an alpine grassland on Qinghai-Tibetan plateau using a two-source model and parameter uncertainty analysis by Bayesian approach. J Hydrol 476:42–51

7

CHAPTER

Multivariate Analysis in Vibrational Spectroscopy of Highly Energetic Materials and Chemical Warfare Agents Simulants

John R. Castro-Suarez, William Ortiz-Rivera, Nataly Galan-Freyle, Amanda Figueroa-Navedo, Leonardo C. Pacheco-Londoño and Samuel P. Hernández-Rivera

[1] ALERT DHS Center of Excellence, Department of ChemistryUniversity of Puerto Rico-Mayagüez, Mayagüez, Puerto Rico

1. INTRODUCTION

The detection of harmful materials in bulk and trace levels present in different matrices: gases/vapors, liquids, and solids is an important consideration for the development of sensors and standoff detection systems for use in National Defense and Security applications. Hazardous chemicals such as highly energetic materials (HEM), homemade explosives (HME), chemical and biological agents are classified as imminent threats, providing terrorists with ways to cause damage to civilians or troops. Chemical warfare agents (CWA) are usually classified as skin-damaging, nerve agents and toxins [1]. Examples of exposures have occurred since World War I with the development of chlorine, phosgene, cyanide and sulfur mustard which were also used in the Iran-Iraq war. In recent times, terrorist attempts involving chemical warfare agents have occurred all over the World as in the case of Sarin (Japan, 1994) and Ricin (London and Paris, 2003) [2]. Threat perception of chemical warfare agents has increased since September 11, 2001 [3,4]. Exposure to low levels of these chemical agents can cause respiratory problems, eye irritation, choking and blisters. LD_{50} (mg/kg) values for Nerve agents include Tabun (0.08), Sarin (0.01), Soman (0.025) and VX (0.007) [5].

Most studies that have been published about detection of these compounds are based on spectroscopic and chromatographic (GC and HPLC) methodologies [6-11]. Vibrational spectroscopy has demonstrated to be valuable for the detection of HEM, HME, CWA and Simulants (CWAS) and Toxic Industrial Compounds (TIC). In particular, infrared spectroscopy (IRS) and Raman spectroscopy (RS) in various modalities have played unique roles in threat compounds detection [6,12-19]. IRS and RS can be employed for detection of Explosives and Chemical Warfare Agents, as well as other chemical and biological threats in airports, in military environments, in government buildings and other public safety places. Raman Spectroscopy employs a non-invasive approach that provides high resolution and specificity. Some applications of IRS and RS includes lab based characterization studies as well as forensic field studies of organic and inorganic substances through their vibration signatures.

Optical Fiber Probes (OFP) have been employed in biomedical applications, in communications, in coupling instrumentation to sensing probes and other important modern applications. Moreover, their uses have been extended to excitation and detection of Raman and infrared signals [16, 20]. Fiber optics applications to Raman Spectroscopy can take advantage of a favorable excitation radiation distribution within the sample; allowing the use of higher laser power levels which, in turn, can yield an elevated signal-to-noise ratio (SNR) for a given experiment without increasing the risk of photo-damaging analytes [21, 22]. In 2011 Ramírez-Cedeño et al. utilized Optical Fiber Coupled Raman Spectroscopy (OFC-RS) to detect hazardous liquids concealed in commercial products [6]. They proved that an optical fiber coupled Raman probe was able to discriminate hazardous liquids inside consumer products from common drinks. Elliason et al. (2007) have also reported drug and liquid explosives detection in concealed in colored plastic containers [23].

Recently infrared spectroscopy has shown progress in the use of more powerful IR sources, such as Quantum Cascade Lasers (QCL) by incorporating these devices in IR reflectance, IR transmission and even in IR microscopy applications [24]. QCL-based setups are being developed for in the field applications such as breath analysis, environmental research, airborne measurements, security applications, laser-based isotope ratio measurements, and many others. In particular, for security applications, optical methods are advantageous because of their capability for remote and standoff detection [14, 25, 26]. Due to improvements in QCL development, mid infrared lasers operating at room temperature with high output powers in the CW regime are commercially available and make it possible to set up a ruggedized system that allows sensing of explosives and others materials outside the laboratory and the ability to enter real world scenarios. With laser based standoff spectroscopy, the detection distance can be a few meters to tens of meters. Because of the inverse square dependence of light intensity, larger distances require high power, collimated light sources such as lasers. For homeland security applications such as detection of suicide bombers or improvised explosive devices, a distance of 50-100 m is generally sufficient.

In this chapter we illustrate the usefulness of incorporating powerful statistical routines to all traditional chemistry disciplines: Chemometrics is the application of statistical tools to plan, execute and analyze experiments in chemistry. To illustrate the power of Chemometrics techniques to analyze experiments in chemistry we have chosen three case studies, all involving identification, quantification, discrimination and classification of chemical threats in different matrices from vibrational spectroscopy multivariate data.

In the first case study a remote Raman detection study was performed for quantification of HEM such as pentaerythritol tetranitrate (PETN) present in different mixtures. The remote measurements were carried out at 10 m by employing a frequency-doubled 532 nm Nd:YAG pulsed laser as excitation line, the quantification study was performed by using partial least squares regression analysis (PLS), Interval-PLS (iPLS) and Synergy-PLS (siPLS) as chemometrics tools to achieve the best correlation between the remote Raman signal and the concentration (%) of PETN explosive in a mixture with pharmaceutical compound.

In the second case study discussed, Optical Fiber Coupled Raman Spectroscopy (OFP-RS) was employed at 488 nm excitation wavelength for detection of a Chemical Warfare Agent Simulant (CWAS): triethyl phosphate (TEP) inside different commercial bottles: green-plastic, green-glass, clear-plastic, clear-glass, amber-glass and white plastic. Aqueous solutions were also used to discriminate on various bottle materials in commercial beverage products.

In a third case study a Fourier Transform infrared interferometer with MCT detector was used for recording vibrational infrared signals from nitroaromatic and peroxide explosives in the gas phase. Furthermore, a dispersive IR HEM detection system using a quantum cascade laser was used to record MIR spectral signals of 2,4,6-trinitrotoluene, pentaerythritol tetranitrate and cyclotrimethylenetrinitramine on travel baggage surfaces. Several models were generated with and without preprocessing throughout MIR spectrum.

2. DESCRIPTION OF METHODOLOGIES

Explosives compounds employed in the studies illustrated were pentaerythritol tetranitrate (PETN) cyclotrimethylenetrinitramine (RDX), triacetone triperoxide (TATP), 2,4-dinitrotoluene (2,4-DNT), 2,4,6-trinitrotoluene (TNT) were synthesized in the laboratory according to methods described by Urbanski [27] and pharmaceutical active compound: acetaminophen (APAP) was purchased from Aldrich-Sigma Chemical Co.

(Milwaukee, WI). Powder mixtures were prepared employing the both compounds (PETN and APAP) mentioned above and the compositions of PETN in the mixtures was varied from 1 to 34% w/w. Components of the mixtures were carefully weighed and mixed using an agate mortar ensuring homogeneity throughout the sample with a total weight of 200 mg.

Figure 1. Commercial bottles (glass and plastic) used to TEP detection using Optical Fiber Coupled Raman spectroscopy.

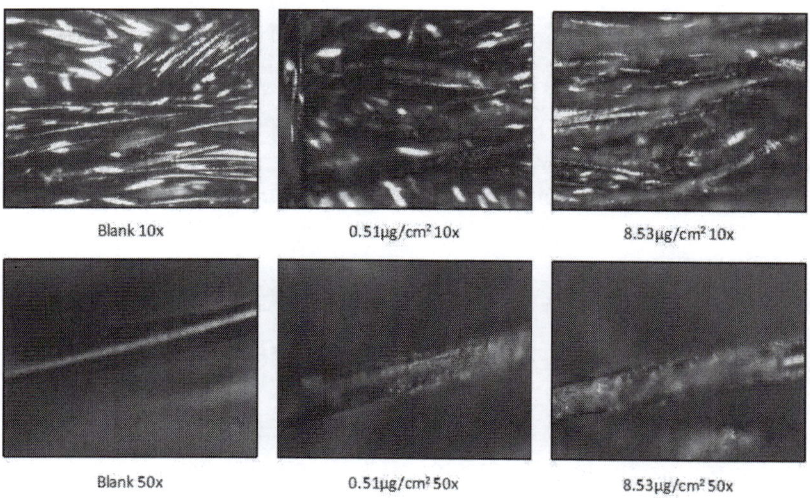

Figure 2. Microscopic Views of PETN on travel baggage substrates; zoom used was 10x and 50x.

A chemical warfare agent simulant: (CWAS) triethyl phosphate 99% (TEP) from Acros Organics (New Jersey, USA) was used to demonstrate the detection and quantification capabilities of this kind of compound by Optical Fibers Coupled Raman spectroscopy. Several formulations were made with the stimulant in their commercial containers: Snapple® Kiwi-Strawberry (Snapple Group USA), Pepsi (PepsiCo Inc., USA), Mountain Dew® (PepsiCo Inc., USA), Heineken® (Mendez & Company, PR), Mott's apple juice® (Mott's LLP., Rye Brook, NY), Leche Suiza® Low Fat Grade A (Suiza Dairy Corp., Puerto Rico) and Malta India® (PR Brewing Co., Mayaguez, PR). The containers can be seen in Figure 1.

For gas phase measurements of 2,4-DNT and TATP using infrared spectroscopy vapors were collected in a gas cell (10 cm long and 3.5 cm diameter) by slowly heating the sample to generate vapors. Finally, traces of PETN, TNT and RDX were placed on travel baggage surface with size of 1 in^2.Figure 2 shows a view of how PETN was deposited on travel baggage substrates at different surface concentration. Figure 3 shows a summary the experiments carried out using vibrational spectroscopy techniques such as infrared and Raman. This figure shows the modalities used, the target tested and the chemometrics models utilized.

For detection of PETN mixed with APAP, Raman spectra of mixtures were acquired by employing a Remote Raman system. This system has been described in detail previously [16]. The prototype was modified using a Headwall Photonics Raman Explorer™ spectrograph with optical layout for 532 nm (Headwall Photonics, Inc.) instead of the Andor spectrometer. The remote Raman system consisted of a MEADE ETX-125 Maksutov-Cassegrain telescope (125 mm clear aperture, 1900 mm focal length f /15). The reflecting collector was coupled to the Raman spectrometer with non-imaging, 200 μm diameter optical fiber (model SR-OPT-8024, Andor Technology, Belfast, Northern Ireland). A frequency-doubled 532 nm Nd:YAG pulsed laser system (Quanta Ray INDI Series, Newport-Spectra Physics, Mountain View, CA) was used as the excitation source. The maximum energy/pulse of the laser at 532 nm was 25 mJ, and it operated at a repetition rate of 10 Hz. The pulse width was approximately 5-8 ns, and the beam divergence was less than 0.5 mrad. A gateable, intensified CCD detector (iStar™ ICCD camera, Model DH-720i-25F-03, Andor Technology, Belfast, Northern Ireland) was used as the photon detector. Andor Technology Solis™ software for spectroscopic, imaging and time-resolved studies was used for spectral data acquisition and processing from the intensified and gated CCD detector. Using this software, the data could be acquired in both imaging and spectroscopic modes. In this experiments, each mixture was placed into a stainless steel sample holder of 0.6 cm in diameter where 30 ft lb of pressure was applied to generate a tablet. Remote Raman spectra of mixtures were measured at a target at telescope standoff distance of 10 m, in the Raman Shift region 450-3000 cm^{-1} using an pulsed laser operating at 532 nm excitation line with a constant energy of 25mJ/pulse (at head) and 100 pulses were applied to achieve spectra with good Signal to Noise. A total of 10 spectra were collected for each mixture acquiring around 56 spectra in the specific Raman shift range.

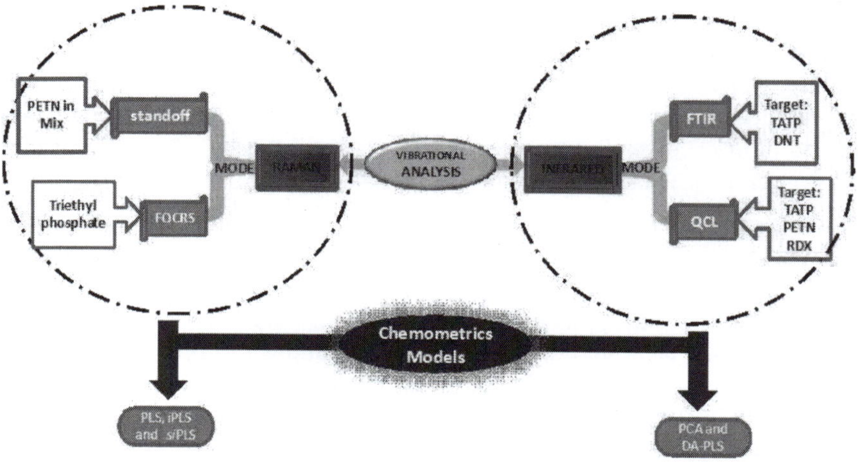

Figure 3. General scheme of experiments using IR and Raman vibrational spectroscopy.

In the detection of TEP chemical warfare simulant in commercial bottles, Raman experiments were performed with a custom built setup (Figure 4) using the strong blue excitation line from an argon ion laser INNOVA 310/8 from Coherent, Inc. at 488.0 nm. The first strand of optical fiber (non-imaging, 600 μm diameter, model AL 1217, Ocean Optics, Inc.) as well as the second (200 μm diameter, model SR-OPT 8024, Andor Technologies Inc.) were used to couple the Raman probe to which a set of laser line filter (to clean satellite lines) and Semrock RazorEdge™ edge filter was used to filter the Rayleigh scattered light. An Andor Technologies spectrograph: Shamrock SR-303i (aperture: f/4; focal length: 303 mm; wavelength resolution: 0.1 nm or 4.2 cm^{-1} at the excitation wavelength) equipped with a 1200 grooves/mm grating was used to analyze the Stokes scattered light. A high performance, back thin illuminated CCD camera (Andor Technologies model # DU970N-UVB) with quantum efficiencies of 90% (200 cm^{-1}) to 95% (3600 cm^{-1}) served as light detector (Figure 4) and calibration was performed using cyclohexane. Detection was performed with a light source in order to be carried out in the high-background environment conditions. In this experiment, six different bottle materials: clear-glass, green-glass, brown-glass, clear-plastic, white-plastic and green-plastic were used to measure the amount of simulant (triethyl phosphate) within the container. Mixtures ranged from 0% to 100% (v/v) of simulant and water, simulant and commercial beverage product were analyzed.

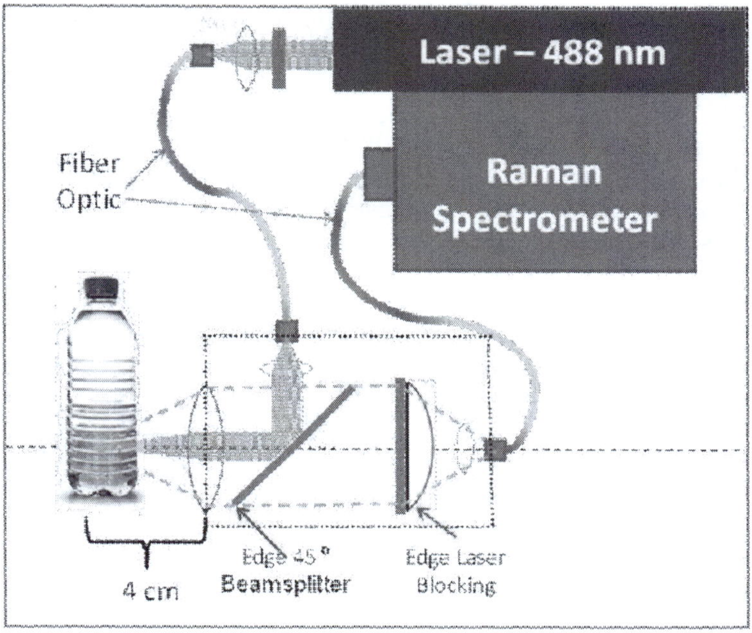

Figure 4. Experiment setup for detection of chemical warfare simulant in commercial bottles.

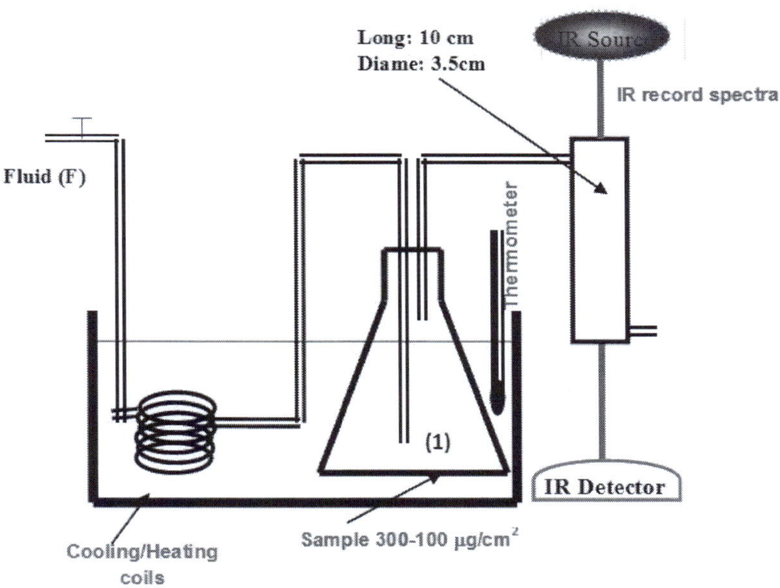

Figure 5. Schematic diagram of experimental setup used in the IR detection. Traces of explosives in gas phase were dragged by air flow and transported to a gas cell for detection.

The IR equipment used for the experiments was a model IFS 66v/S interferometer (Bruker Optics, Billerica, MA). For the experiments described, the system was equipped with a DTGS detector and a potassium bromide (KBr) beamsplitter. A spectroscopic measurement averaged 20 scans at a resolution of 4 cm^{-1} using OPUS Version 4.2 software in the range of 7500 – 400 cm^{-1}. Target chemicals were TATP, 2,4-DNT and 2,4,6-TNT. Second, an EM-27 Open Path FT-IR interferometer (Bruker Optics, Billerica, MA) was used to obtain the IR spectral information of gas phase TATP samples with a (TE) cooled MCT detector. Third: Quantum Cascade Laser (QCL) based dispersive IR spectrometer LaserTune™ (Block Engineering, Marlborough, MA) was used to obtain the IR spectral information of TATP samples. An Agilent 6890 gas chromatograph (GC) coupled to an Agilent 5893 mass selective detector (MSD) with a capillary column: HP-5 MS 5% phenyl methyl siloxane, Length: 30.0 m, 250.0 µm in diameter and 0.25 µm of film thickness was used for detecting the presence of explosive TATP, 2,4-DNT and 2,4,6-TNT in the gas phase.

Figure 5 shows a schematic diagram of the experimental setup used in the investigation. Samples of 100-300 $\mu g/cm^2$ of explosives were placed on the bottom of a 500 mL Erlenmeyer flask at the position labeled (1) in Figure 5. A flow of dry air (1-16 mL/s) at temperatures of 0-38°C was used to transfer the solid explosives to the gas phase. The measurements as function of temperature were done in two forms, scanning the range of temperature and fixed temperature point measurements. Traces of explosives in gas phase were dragged from the surface by the air flow and transported to an IR gas cell for detection. Spectra were recorded using the instrument first at 4 cm^{-1} of resolution and 25 scans were used for the experiments. A total of 1089 spectra of air with 2,4-DNT, 1194 spectra of air with TATP and 2200 spectra of air were recorded to generate the models. On the OP EM-27 active mode experiments were carried out at lab temperature (25°C) at 30 scans and 4 cm^{-1} resolution. The spectral range used was from 700 to 1600 cm^{-1}. Experiments using QCL LaserTune™ active mode experiments were carried out at lab temperature of 20 °C at 1 spectrum and 4 cm^{-1} resolution. The spectral range used was from 830 to 1430 cm^{-1}. The presence of TATP in air was determined by GC-MS, and the concentration of 2,4-DNT in air for different flow conditions and temperatures were calculated by calibration curves from GC-µECD. Finally, explosive traces (PETN, TNT and RDX) were deposited on travel baggage of 1 in^2 size of and analyzed with the QCL spectrometer.

3. CHEMOMETRICS TO VIBRATIONAL SPECTRAL ANALYSIS

The automation and computerization of laboratories have been carried out with various important consequences. One of them is the rapid acquisition of large amounts of data. However, it is well know that acquiring such large amount of data is far from to providing appropriate answers quicker. Obtaining vibrational spectroscopy multivariate data is not synonymous with possessing vibrational information. The later must be interpreted and placed in context to convert it into useful information for the user. Chemometrics is the field of chemistry that provides the user with the required tools to enable that capability.

A great deal of chemometrics tools have been developed and tested. The most used of these tools to identify, quantify and classify data sets are those that make use of principal components analysis (PCA), partial least squares (PLS), discriminant analysis (DA), and their combined usage: PLS-DA and hierarchical cluster analysis (HCA). PCA transforms a set of variables into fewer variables (called factors, rank, dimensions or principal components) which contain most of the information (variance) of the initial data set [28-31].

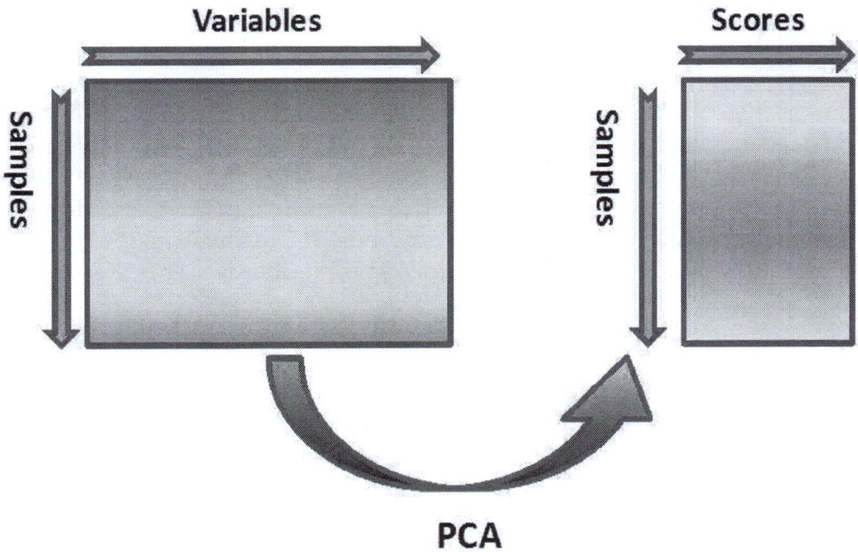

Figure 6. Simplified scheme for a PCA analysis.

The PCA algorithm seeks to save the information from a large number of variables in a small number of uncorrelated components, with minimal loss of information. The main reasons for performing a PCA are: reduction of the number of variables to fewer dimensions that contain as much information as possible and have uncorrelated dimensions (used to avoid multi-collinearity in multiple regressions, among other things). An important method for qualitative analysis of spectral data is principal component analysis. PCA is a method for the investigation of the variation within a multivariable data set. The first step in PCA is to subtract the average value or spectrum from the entire data set, this is called mean centering. The largest source of variation in the data set is called principal component PC1. The 2nd largest source of variation in the data, which is independent of PC1, is called PC2. Principal components form a set of orthogonal vectors. For each one of the data points, the projection of the data point onto the P1 or P2 vector is called a score value. Plots of sample score values for different principal components, typically P1 versus P2 are called score plots. Score plots provide important information about how different samples are related. Principal component plots, also called loading plots, provide information about how different variables are related to each other. In practical cases, PCA uses a single X matrix which is represented by the infrared spectra. PCA is a purely qualitative analysis (does not give a quantitative value that establishes how different are a spectral dataset) to visualize if there is variability between a set of IR spectra. PCA can thus also be used to detect the presence of outliers. Figure 6 shows a simplified PCA scheme [28-29].

Partial least squares (PLS) regression is a quantitative spectral decomposition technique that is closely related to PCA regression. The importance of PLS is that it is used to design and build robust calibration models for multivariate quantitative analysis. PLS actually uses the concentration information during the decomposition process. This causes spectra containing higher constituent concentrations to be weighted more heavily than those with lower concentrations. The main idea of PLS is to get as much concentration information as possible into the first few loading vectors (number of component, factors, ranks or principal components). PLS regression consists of two fundamental steps. First, to transform the ***X*** predictive matrix (spectra) of order n × p (n = number of samples and p = number of variables: cm^{-1} or nm), in an matrix of components or latent variables uncorrelated, ***T*** = (T1,..., Tp) of order n × p, called PLS components, using the ***Y*** response vector (concentrations) of order n × 1; this contrasts with the principal component analysis in which the components are obtained using only the ***X*** predictive matrix. Second, to calculate the estimated regression model using the ***Y*** response original vector as predictive, PLS components. The dimensionality reduction can be applied directly on the components as they are orthogonal. The number of PC required for the regression analysis must be much smaller than the number of predictors. There is a number of ways of expressing these, a convenient one being (Eq.1 and 2) [29]:

$$\boldsymbol{X} = \boldsymbol{T}.\boldsymbol{P} + \boldsymbol{E} \quad (1)$$

$$\boldsymbol{c} = \boldsymbol{T}.\boldsymbol{q} + \boldsymbol{f} \quad (2)$$

Figure 7 illustrates a simplified scheme for PLS: ***X*** represents the experimental measurements (e.g. spectra) and ***c*** (or ***Y***) the concentrations.

The first equation above appears similar to that of PCA, but the scores matrix also models the concentrations, and the vector ***q*** has some analogy to a loadings vector. The matrix ***T*** is common to both equations. ***E*** is an error matrix for the ***X*** block and ***f*** an error vector for the ***c*** block. The scores are orthogonal, but the loadings (***P***) are not orthogonal, unlike in PCA, and usually they are not normalized.

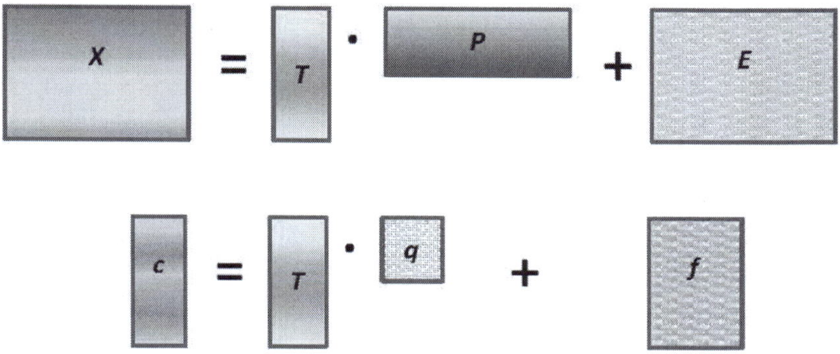

Figure 7. Simplified scheme for a PLS transformation.

Chemometrics techniques have improved the last years in order to save time and computational resources in different models to be used without compromising the quality of results. In 2000 Norgaard and co-workers [32,33], developed different algorithms useful in Chemometrics field called interval partial least squares (iPLS) and this tool was presented for use on NIR spectral data. Recently, this new graphically oriented local modeling procedure has been implemented in many areas of research such as petrochemicals, pharmaceutical and beverage industry [34-36].

The principle of the iPLS is to optimize the predictive capability of PLS regression models and to support in interpretation. This algorithm which develops local PLS models on equidistant subintervals of the full-spectrum region. Its major objective is to provide an overall perspective of the significant information in different spectral subdivisions, thereby focusing on important spectral regions and removing interferences from other regions. The sensitivity of the PLS algorithm to noisy variables is highlighted by the informative *i*PLS plots [32]. For synergy interval PLS (siPLS), the basic principle of this algorithm is the same as iPLS first, it is to split the data set into a number of intervals (variable-wise), next, to develop PLS regression models for all possible combinations of two, three or four intervals. Thereafter, RMSECV is calculated for every combination of intervals. The combination of intervals with the lowest root mean square error of cross-validation (RMSECV) is selected.

Finally, cluster analysis is the name given to a set of techniques that seeks to determine the structural characteristics of multivariate data sets by dividing the data into groups, clusters, or hierarchies. For cluster analysis, each sample is treated as a point in an n-dimensional measurement space. The coordinate axes of this space are defined by the measurements used to characterize the samples. Cluster analysis assesses the similarity between samples by measuring the distances between the points in the measurement space. Samples that are similar will lie close to one another, whereas dissimilar samples are distant from each other [28].

In this chapter, remote Raman detection experiments were performed to quantify HEM such as PETN present in mixtures with non-HEM. The remote measurements were carried out at 10 m employing a frequency-doubled 532 nm Nd:YAG pulsed laser as excitation source. The quantification study was performed by using PLS, iPLS and siPLS as chemometrics tools to achieve the best correlation between the remote Raman signal and the concentration (%) of PETN explosive in a mixture with pharmaceutical compound. Discrimination of chemical warfare agent simulant (CWAS) TEP concealed within commercial beverage bottles using Optical Fiber Coupled Raman Spectroscopy with the use of different chemometrics techniques such as PLS, PLS-DA. Finally infrared spectroscopic information analysis using Chemometrics was designed and implemented in the detection of HEM: 2,4-DNT, TATP, PETN and RDX, present at trace level on surfaces and in air were analyzed by Chemometrics Enhanced Vibrational Spectroscopy.

4. MULTIVARIATE DETECTION AND QUANTIFICATION FROM VIBRATIONAL SPECTRA

4.1. Remote Raman Experiments

Different preprocessing methods such as vector normalization (VN), mean centering (MC), auto scaling (AS), multiple scattering correction (MSC), standard normal variate (SNV) and first and second derivatives have been developed to improve a good multivariate quantification. The 56 remote Raman spectra taken from PETN detection in mixes with APAP were randomly split into two groups: a first group with the 70% of the data for calibration and cross validation (training set) and a second group for external validation (test set) formed by the remaining 30% of the data. The quantitative model was performed by using chemometrics tools, such as PLS, iPLS and siPLS. The PLS program used was from PLS-ToolBox™ (Eigenvector Research Inc.) for use with MatLab™. The iPLS and siPLS algorithms used in this work were carried out by employing iToolbox™, (downloaded from http://www. models.kvl.dk). The performance of the final PLS, iPLS and siPLS models were evaluated according to the root mean square error of cross-validation (RMSECV), a leave-one-sample-out cross-validation method and the predictive ability of the models were assessed by the root mean square error of prediction (RMSEP) and the correlation coefficient (R). In general for all the PLS models RMSECV were calculated as follows:

$$RMSECV = \sqrt{\frac{\sum_{i=1}^{n_{cal}{}^{l}} (c_p - c_i)^2}{n_{cal}}} \quad (3)$$

Where c_i and c_p are the experimental and predicted concentration, respectively, of the ith calibration sample when situated in a left out segment, n_{cal} is the number of calibration samples in the training set. The number of PLS components included in the model is selected according to the lowest RMSECV. This procedure is repeated for each of the preprocessed spectra. For the test set, the root mean square error of prediction (RMSEP) is calculated as follows:

$$RMSEP = \sqrt{\frac{\sum_{i=1}^{n_{test}} (c_i - c_p)^2}{n_{test}}} \quad (4)$$

The best model with the overall lowest RMSECV will be selected as final model. Correlation coefficients between the predicted and the true concentration are calculated for both the calibration and the test set, which are calculated as follows from Equation 5, where Ci–Ci- is the mean of the experimental measurement results for all samples in the train and test sets.

$$R = \sqrt{1 - \frac{\sum_{i=1}^{n} (c_p - c_i)^2}{\sum_{i=1}^{n} (c_i - \bar{c_i})^2}} \quad (5)$$

The implementation of new methodologies for enhanced detection of hazardous compounds such as explosives is always attractive for many countries principally for defense and security applications. Terrorist employ different ways to pose threats and make illegal acts against military and civilian people. According to this situation our study is focused on detection of explosives present in mixture prepared intentionally with a pharmaceutical product by employing remote Raman detection and chemometrics tools. Remote Raman spectra of PETN, APAP in mixtures of them are illustrated inFigure 8. The results show that mean centering (MC) pre-processing method was the most successful method for correcting background and was selected for construction of further models because they presented small improvement in RMSEC.

The full spectrum was split in 20 independent intervals and the RMSECV values for PLS models constructed with different intervals is shown in Figure 9. Models with no intervals were better than PLS models with all variables (dotted in line) and the intervals 6 (1185.2-1328.9 cm^{-1}), 9 (1619.8 -1755.4 cm^{-1}), and 19 (2878 -2988.4 cm^{-1}), presened the lowest

RMSECV values where more variability exists. These values are shown in Table 1. The number of latent variables required for the models obtained using different intervals is the numbers shown inside the rectangles.

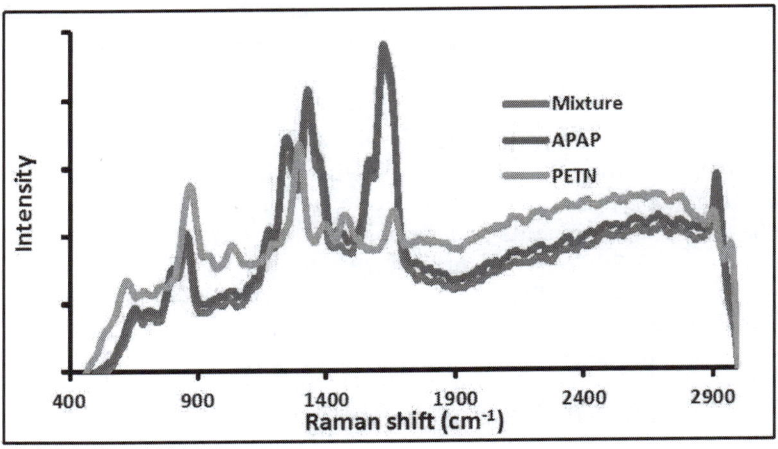

FIGURE 8. Remote Raman spectra of PETN, APAP and mixture of them, collected at 10 m of target to collector distance employing 532 nm laser with 100 pulses of 25 mJ/pulse.

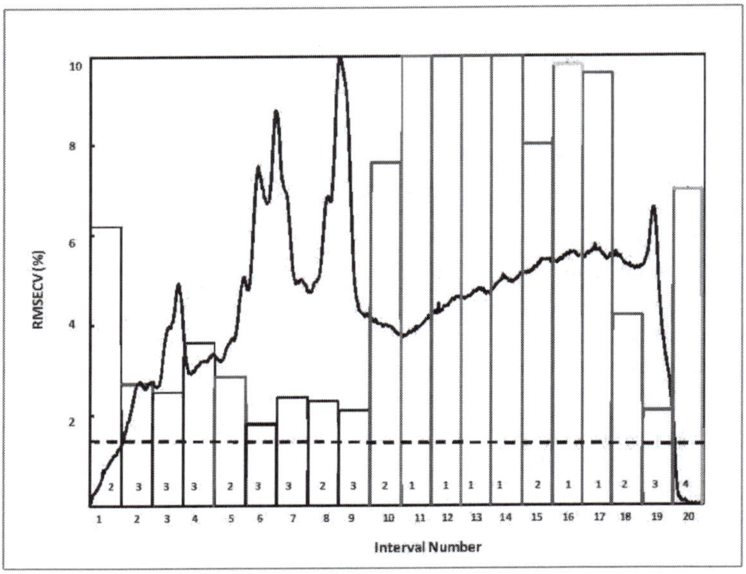

Figure 9. RMSECV values for PLS models obtained for the 20 different intervals (bars) used in iPLS models. The horizontal dotted line represents the RMSECV value for the PLS model with all variables. Numbers inside the rectangles are the optimal number of latent variables.

TABLE 1. Full-cross-validated PLS, iPLS, and siPLS models for prediction of PETN in the range 1.0–34.0% Remote Raman spectra. All models are based on MC data.

Models	LV[a]	Intervals	NV[b]	RMSECV (%)	RMSEP (%)	R_{CV}	R_P
PLS	6	All	730	1.8	2.2	0.978	0.979
iPLS	3	6	37	2.0	2.8	0.986	0.976
	3	9	37	2.5	3.1	0.976	0.969
	3	19	36	2.7	2.4	0.972	0.988
siPLS	7	3,9,19	110	1.4	1.8	0.993	0.992

a Latent variable b Total number of variables.

In synergy interval-PLS (siPLS) model calibration, the number of intervals was also optimized according RMSECV values. Table 1 shows the results of siPLS model calibration when the spectra were split into different number of intervals. The optimum siPLS model was obtained with the combination of 3 intervals (3, 9 and 19) and 7 PLS components. The lowest RMSECV was 1.4, compared with RMSECV values obtained for PLS model with all variables and iPLS models. According to the statistical results illustrated in Table 1, it is important to establish that iPLS or siPLS models with 4 or more intervals (data not shown) including intervals 10-18 were explored. These intervals correspond to noisy areas which were not eliminated in order that the models could choose the spectral region of larger variability. The capability of prediction of siPLS models was better when compared to the other models As shown in the correlation plots in Figure 10, there is a good relationship between the True and Predicted concentration (%) for PETN, with R_{CV} values of 0.993. This can also be appreciated by the good prediction of the test set of samples with values of RMSEP of 1.8% for the corresponding explosives. The final model separated the vibrational spectra into 20 intervals, 7 latent variables were used and the intervals number 3, 9 and 19 were combined The selected intervals included regions of 724.2 - 876.7cm^{-1},1619.8-1755.4 cm^{-1} and 2878 -2988.4 cm^{-1}, The first Raman shift region correspond to NO_2 scissoring mode and O-N stretching band; the second region is relevant for NO_2 asymmetric stretching mode and C=O stretching band; the third region represents the C-H stretching mode [37-39].

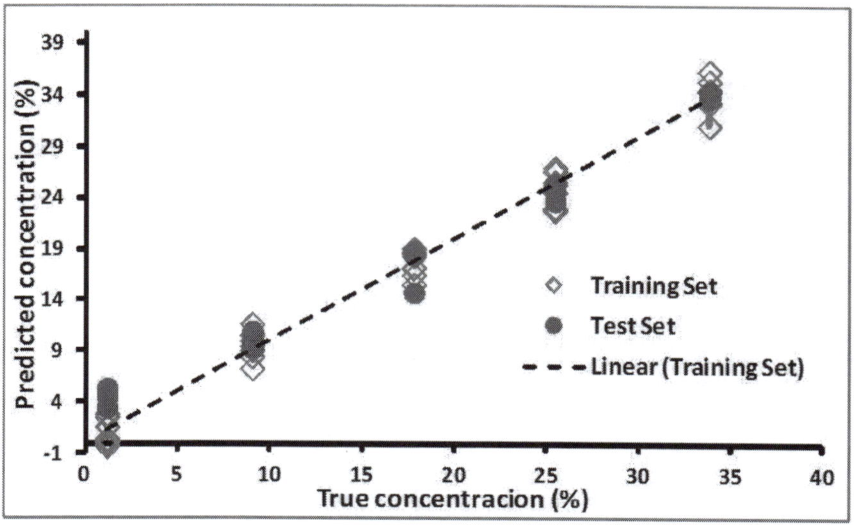

Figure 10. Predicted vs. True PETN concentration for siPLS model using 3, 9, and 19, intervals and 7 latent variables.

4.2. Optical Fiber Probe Raman Spectroscopy Experiments

In the optical fibers coupled Raman spectrum of TEP, shown in Figure 11, the CWAS has characteristics peaks at 733 cm^{-1} (PO_3 symmetric stretching mode), 813 cm^{-1} (PO_3 asymmetric stretch), 1032 and 1098 cm^{-1} (C–O stretch), 1162 cm^{-1} (CH_3 rocking) and 1279 cm^{-1} (P–O symmetric stretch) [6]. Mixtures of TEP with commercial liquids were measured in their corresponding commercials bottles. TEP concentration varied from 0 to 100 (%v/v). In Figure 12, TEP Raman spectra are shown for different bottle materials. At all concentrations, the TEP characteristic peaks could be distinguished within the different types of materials of the container with the exception of brown glass and white plastic. These two bottle materials had lower transmittance in the 200 to 1400 cm^{-1} region and TEP characteristic peaks in the 2700 to 3200 cm^{-1}region. UV-VIS spectra (data not shown) show the increased absorbance in bottle materials such as white plastic and amber glass (Malta™). This confirms nature of the low intensity Raman peaks in the region (200-1400 cm^{-1}) shown in Figure 12. When light scatters turbid materials, such as amber glass or white plastic, the material is absorbing or blocking the light when compared to clear glass and clear plastic. Thickness of the bottle material and coloration also play a role in absorbance and transmission. The high intensity peak at 2300 cm^{-1}corresponds to the background light (mercury vapor from fluorescent lamps). This peak is shown with higher intensity in Raman spectra of brown glass and white plastic in comparison to the rest of spectra due to the increase in integration time for these two bottle materials. All bottle materials were subject to background light in order to simulate real-time conditions found in military, airport and other environments where a light source is involved. This analysis is based on increased absorbance shown in the UV-VIS Spectra for different bottle materials (data not shown).

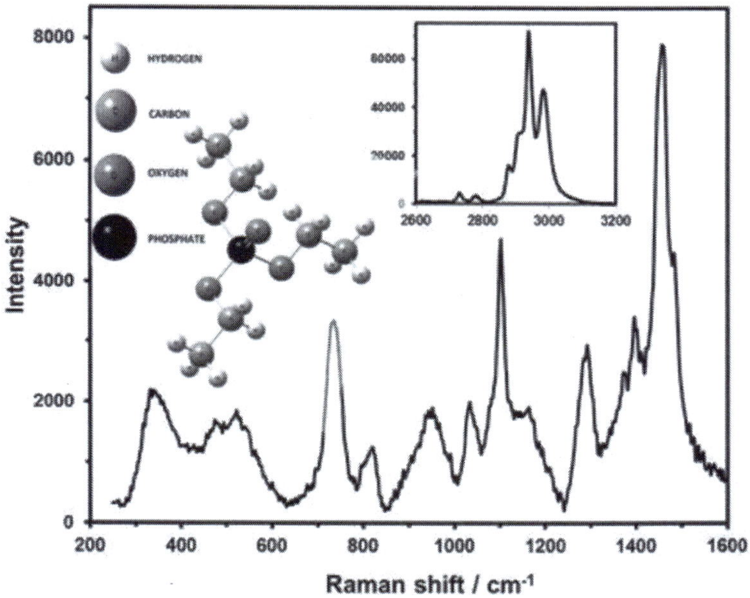

Figure 11. Raman vibrational spectrum of TEP excited at 488 nm.

Calibration models were performed with PLS regression model to distinguish between the samples that contain TEP in aqueous solutions compared to the solutions with TEP and the commercial product. In Figure 13, eight PLS regression models are chosen in order to show the marked difference between the best and the worst regression model, each of these with and without pre-processing steps. Since integration times were normalized for each bottle, Limits of Detection were similar with the exception of amber/brown glass (Malta™). The aqueous solutions show a better R^2 values than the mixtures with the commercial product. For clear glass (Snapple™), the R^2 value is 0.9925 for aqueous solutions compared to 0.9747 for the mixtures with the commercial product. The R^2 values for Malta™ in aqueous solutions showed a significant increase with optimization (0.4193 without preprocessing and 0.9508 with optimization). However, optimization with Malta™ shows a lower R^2 value 0.7646 compared to 0.8047 without preprocessing.

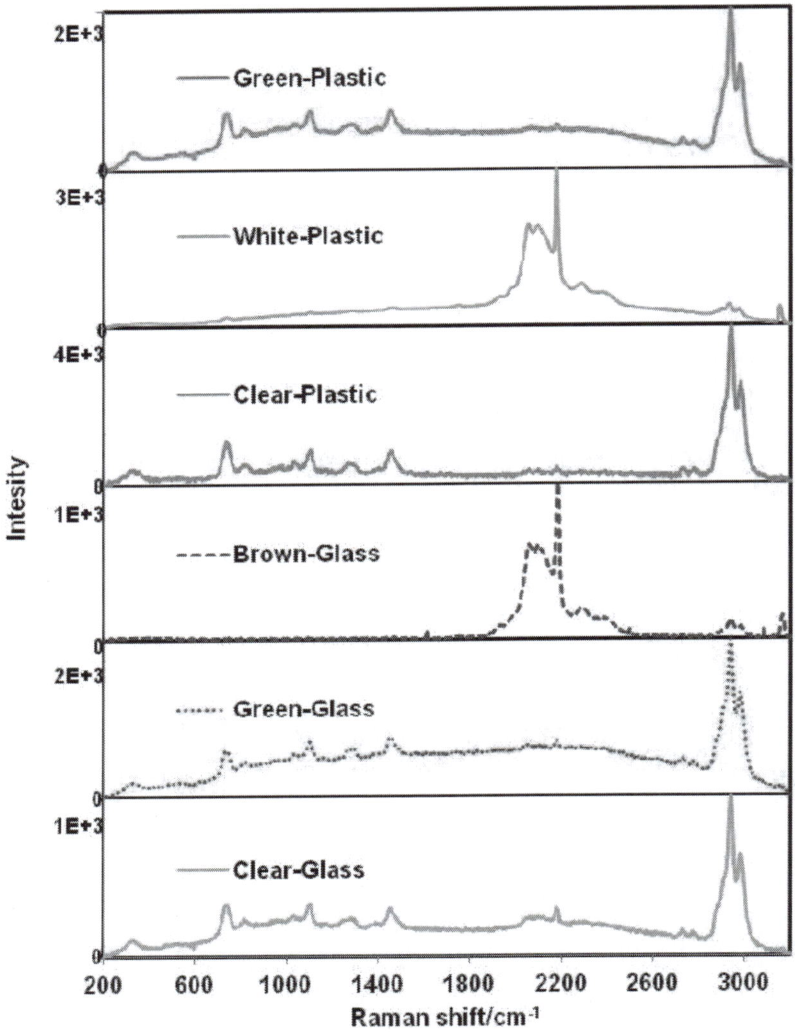

Figure 12. Raman spectra of Triethyl Phosphate in various types of bottle materials

It is clear that the R^2 values increase in PLS regression models for aqueous solutions since water does not present strong signatures in Raman Spectroscopy. Every other PLS regression model (green plastic, green glass, clear plastic, clear glass, and white plastic) in aqueous and beverage solution presented nearly similar limits of detection. Each of these limits improved with their respective preprocessing step (vector normalization, standard normal variate and mean centering). With the help of integration time for each bottle material, normalization was achieved with the limits of

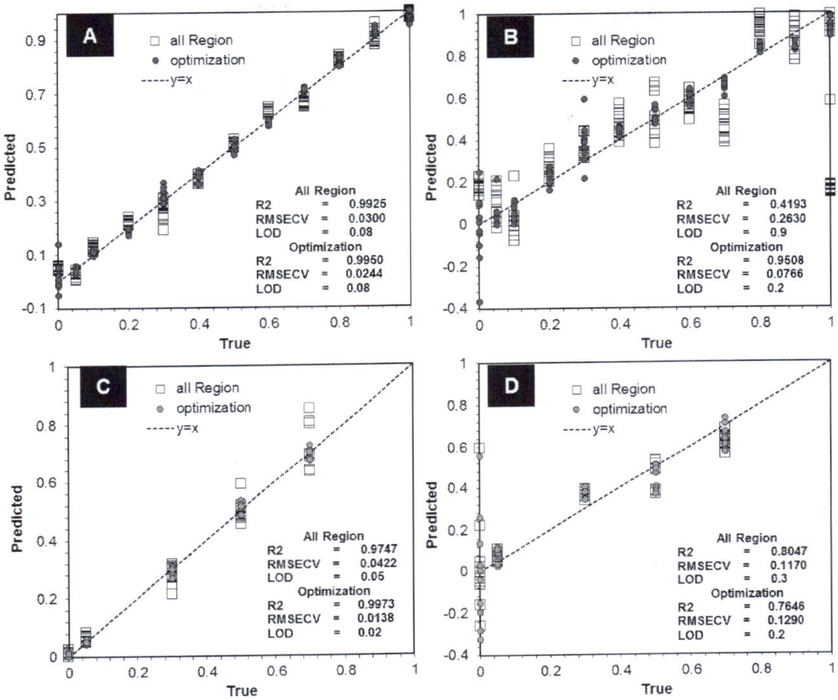

Figure 13. A) PLS models of TEP in aqueous solution in Snapple™ container (clear glass materials) with (vector normalization) and without preprocessing. B) PLS models of TEP in aqueous solution inside Malta™ container (amber glass materials) with (mean centering, standard normal variate) and without preprocessing. C) PLS models of TEP mixtures with the commercial product Snapple™ (clear glass materials) with (vector normalization) and without preprocessing. D) PLS models of TEP mixtures with the commercial product Malta™ (amber glass materials) with (mean centering) and without preprocessing.

detection and root-mean-squared error cross-validation (RMSECV). These values were found as acceptable in an average between the best models of approximately 2.5%. The Limits of Detection for PLS methods were estimated using the equation 6 [40]:

$$LOD=\Delta\ (\alpha,\beta,\upsilon)\ \mathrm{x}\ RMSEC(\sqrt{1}+h_0) \quad (6)$$

Root mean squared error of calibration (RMSEC) was obtained from the square fit errors $[(c_{predicted} - c_{true})2/\upsilon]^{1/2}$ where the sum extends to all samples of the calibration set. The degrees of freedom were then calculated as $\upsilon = n - F - 1$ where F is the number of latent variables and n is the number of samples in the set. The distance of the predicted sample from zero concentration to the calibration set's mean is the leverage h_0. Ultimately, $\Delta(\alpha,\beta,\upsilon)$ corresponds to a statistical parameter that notices the α and β probabilities of falsely stating presence/absence of the chemical warfare agent stimulant. Since $\upsilon \geq 25$, we used $\alpha = 3.4$ for the LOD. LOQ values as per Eq. 7 were studied at a concentration with a Relative Standard Deviation (RSD) of 15% as stated by Felipe-Sotelo *et al.* [40]:

$$\mathrm{LOQ} = 100\mathrm{x}(\mathrm{RMSECx}(1 + \mathrm{h_0})^{0.5}\mathrm{RSD}(\%) \quad (7)$$

Comparing limits of detection (Figure 13) the same integration times were used for aqueous and commercial beverage bottle solutions. A and B (Figure 5) show Snapple™ and Malta™ in aqueous solutions with TEP. Figures 13C and 13D in the same figure show mixtures of TEP with commercial products and with less data (5, 30, 70 and 0 %v/v) due to limited time. Snapple has lower limits of detection which is favorable for detection of chemical warfare stimulants in commercial bottles made out of various materials. When comparing limits of detection for aqueous solutions versus solutions with commercial beverage product inside commercial bottles, limits of detection are considerably lower. R^2 prediction values were higher in aqueous solutions since water does not present significant Raman signal. Limits of detection were found as low as 1 percent for white plastic. Optimization also improves or lowers the limits of detection as shown in Figure 13.

Table 2 shows Limits of Detection and Quantification (LOD and LOQ respectively) for various commercial beverage bottle solutions with TEP for the best models. Preprocessing options include Vector Normalization (V.N.), No preprocessing (N/A), Mean Centering (M.C.), Constant Offset Normalization (C.O.N.), First Derivative (F.D.) and Multiplicative Scatter Correction (M.S.C). Higher limits of detection and quantification for amber glass and clear plastic were presented due to their dark coloration in bottle material (amber) and commercial beverage product (Pepsi and Malta). An unexpected low value for limits of detection and quantification for white plastic was observed. This may be due to the low amount of trials (5 instead of 10 for 5, 30, 50 and 70 (%v/v of TEP) as was done with other bottle materials due to the high integration times for this material. Even though TEP, a surfactant agent, did not present a homogeneous solution with milk, integration times were normalized in order to obtain a better model of a clear linear regression with an R^2 value of 0.9987 and excellent limits of detection of 0.01(1%).

4.3. Gas Phase Infrered Spectroscopy Experiments

Multivariate calibration methods such as Partial Least Squares (PLS) models can be formulated as a regression equation [41, 42]. The equation in metrical form is **Y** = **XB**, where **B** is computed as **B** = **W(PTW)**$^{-1}$ **QT** and **W** is the matrix of weights of **X**, **Q** is the loadings matrix of **Y**, and **P** is the **X**loadings matrix. In this study, the **Y** matrix represents the dependent variables but this is changed from continuous to discrete variation, and contains information about different classes of objects [43-45], it is a simple two states function: 1 represents the condition for the presence of explosive in the sample and 0 stands for the absence of explosive in the sample analyzed. By these means it will be possible to decide if an explosive substance is present or not in a sample. The values originating from the analysis: wavenumber range or parts of spectra are the independent variables (**X** matrix). In this study the loading vectors or number of component (**B** matrix) were used for independent variables in the DA.

Table 2. Limits of Detection and Quantification for the PLS models of TEP in commercial beverage bottles and aqueous mixtures along with their respective preprocessing methods.

	COMMERCIAL BEVERAGE MIXTURES					
	Green Glass	White Plastic	Amber Glass	Clear Glass	Clear Plastic	Green Plastic
LOD (%)	3	1	26	4	22	3
LOQ (%)	8	3	77	11	66	9
Preprocess	V.N.	V.N.	N/A	M.C. + V.N.	C.O.N.	M.S.C. + M.C.
	AQUEOUS MIXTURES					
LOD (%)	11	7	16	8	8	4
LOQ (%)	33	21	48	22	25	12
Preprocess	V.N.	F.D. + V.N.	F.D. + V.N.	V.N.	V.N.	N/A

TATP and 2,4-DNT in air were detected using FTIR spectroscopy. At trace levels, the vibrational signatures are not easily perceptible. Vibrational signatures of explosive can be confused with vibrations arising from the background air components. Thus the first task was to determine the possible interference of the two spectra. Figures 14a and 14b show the spectra of flowing gas that contains TATP and 2,4-DNT traces. The characteristic infrared signals of TAPT at 1200 cm^{-1} and at 1550 cm^{-1} for 2,4-DNT can be observed in Figure 14 which confirms the presence of these compounds in air. Linear Combination Analysis in the form of Partial Least Squares (PLS) was calculated for all FTIR spectra (7500-600 cm^{-1}). Two and four vectors were required to find the perturbation produced by TATP and 2,4-DNT respectively, on the normal flowing air IR spectrum. The discriminating function used was a two position switch type function: On – Off (Yes/No). The nomenclature in the DA was for classification of samples in terms of "Disc-1=TATP present" or "Disc-0 = TATP not present" in air, for TATP; and "Disc-1 = 2,4-DNT present" or "Disc-0 = 2,4-DNT not present" in air, for 2,4-DNT. The results were presented in the form of histogram, where the y-axis is the frequency and x-axis is the discrimination function. Also the prediction of new sample was present in this form, (Figures 15 and16) in these graphs, the improvement of models, when vectors are added successively is observed.

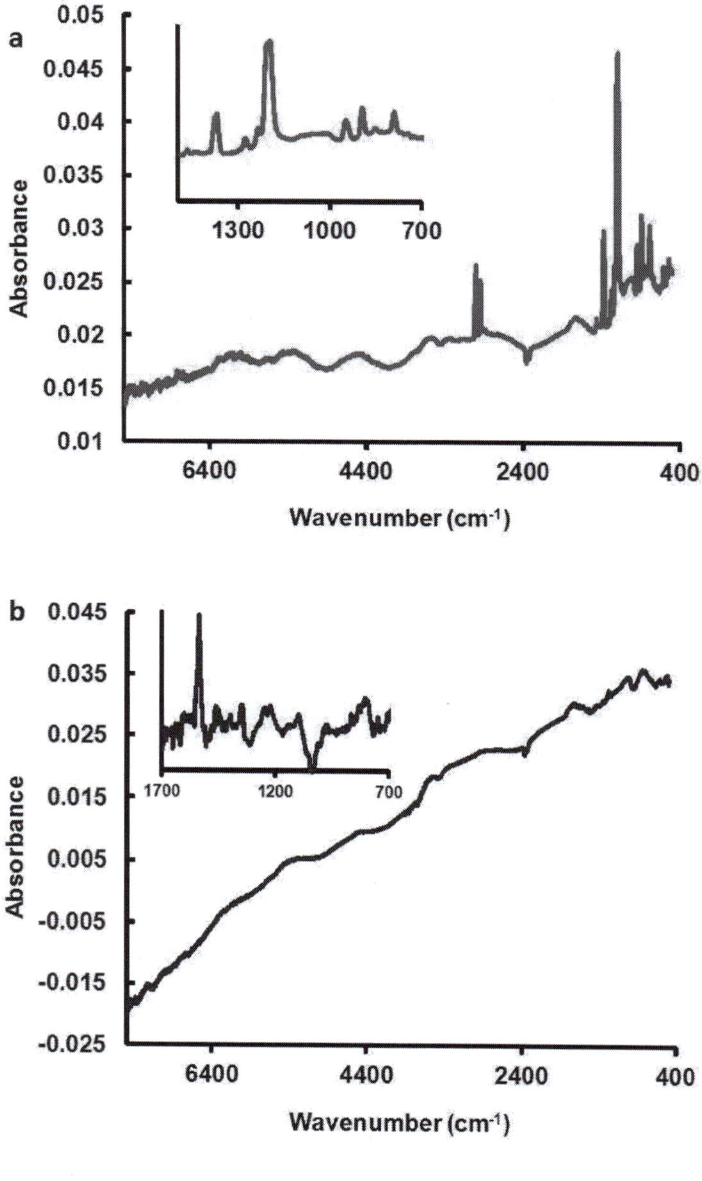

Figure 14. FTIR vibrational spectra of gas explosive in air: a. TATP and b. 2,4-DNT traces.

The best discriminant function was selected based on statistical significance (p) and the percentage of cases correctly classified (PCCC) [46]. The validation was done by internal jackknifing validation and external validation. Internal validation: in this method, each spectrum was successively removed from the data set, and then it was discriminated from a new model built from the remaining spectra. This procedure was done for each one of the spectra in the data set, and the predicted discriminations were then compared with the experimental observations. The generated percentage of cases correctly classified is called the cross-percentage of cases correctly classified (PCCCC). External validation: before making the model, 100 spectra of air, 100 spectra air with TATP and 50 spectra air with 2,4-DNT were taken from the data set randomly. These spectra were analyzed by the validation model.

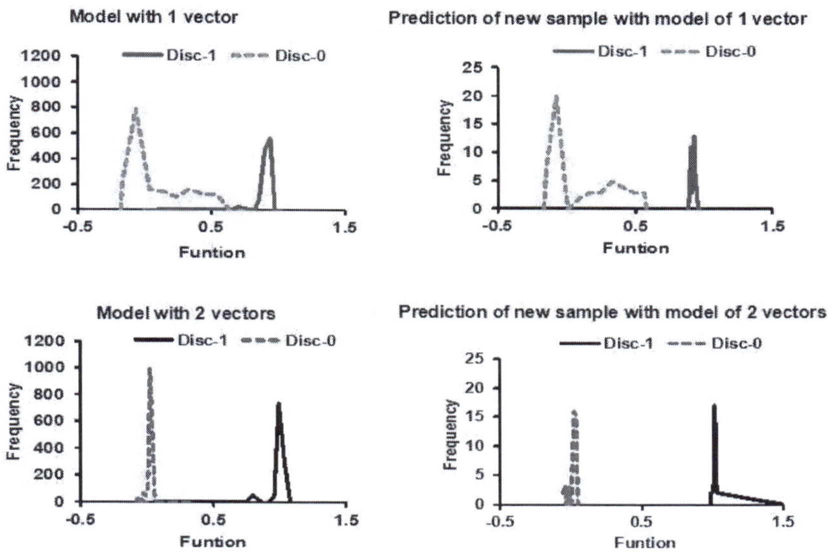

Figure 15. Histogram for discrimination models of TATP and external validation.

For the PCA analysis of TATP in air spectra were recorded using the EM-27 and the LaserScope™ instruments. A total of 60 spectra were recorded from clean air and 120 spectra from air with TATP present using EM-27 and 35 spectra were recorded from clean air and 30 spectra from TATP present in air using LaserScope™. All PCA analysis including any preprocessing in the spectral data were run using PLS-Toolbox software. PCA runs were made with the raw data and using different preprocessing treatments. The preprocessing treatments used were: auto scale, smoothing, SNV-standard normal variation, Mean center, auto scale + 1st derivate, auto scale + 2nd

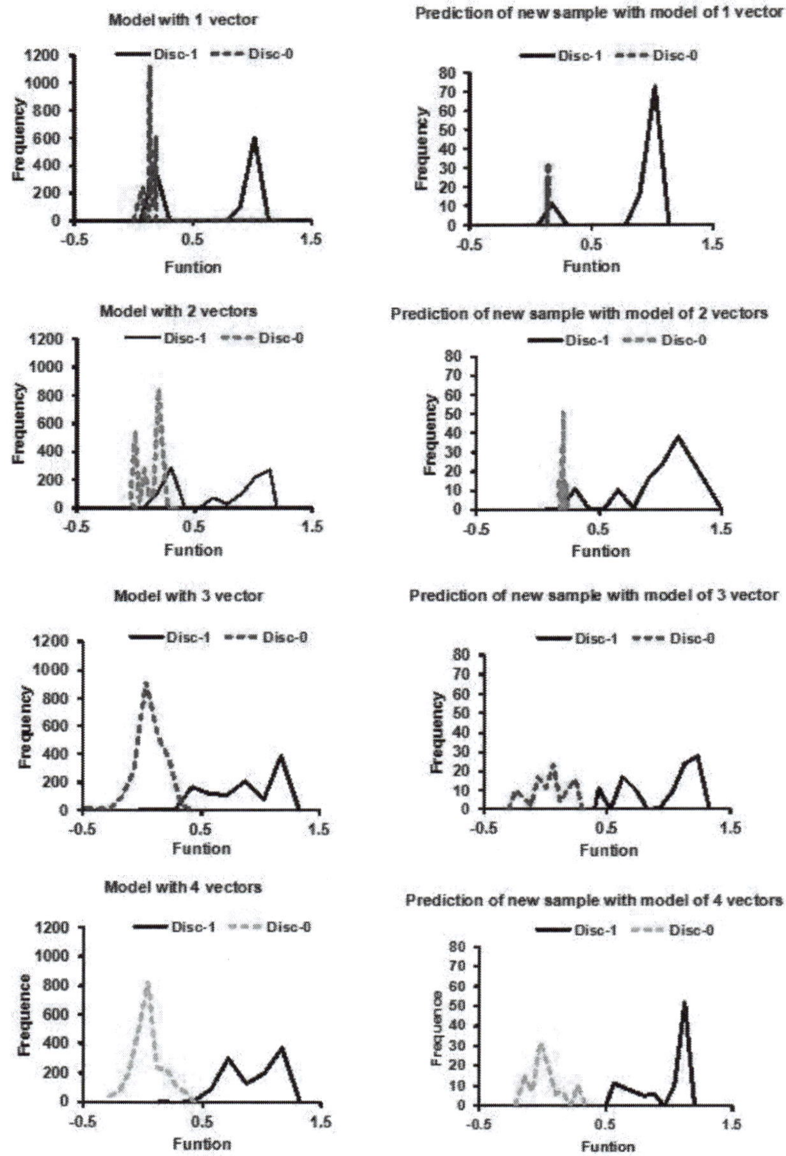

Figure 16. Histogram for discrimination models of 2,4-DNT and external validation.

derivate, mean center + 1st derivate, mean center + 2nd derivative, MSC-multiplicative scattering correction. The algorithm used to carryout smoothing and derivatives was that ofSavitzky-Golay (every 11, 17, 21 and 31 points). During the PCA runs it was not necessary to eliminate spectral data.

The infrared data from clean air and air with TATP were run together in the PCA model for each instrument used. Figure 17 shows the Scores plots for the PCA obtained. Figure 17a shows the first two principal components from spectral data using the EM-27 FTIR spectrometer with a Globar sourceFigure 17b shows the first two principal component analyses from spectral data of TATP detection from LaserScope™ spectrometer using Quantum Cascade Laser Source. The best results achieved for both PCA models (illustrated in Figures 17a and 17b) were using raw data. Both results allowed classifying gas phase TATP explosive from clean air. In Figure 17 can be noticed that PC1 tends to relate the differences between the IR dataset two.

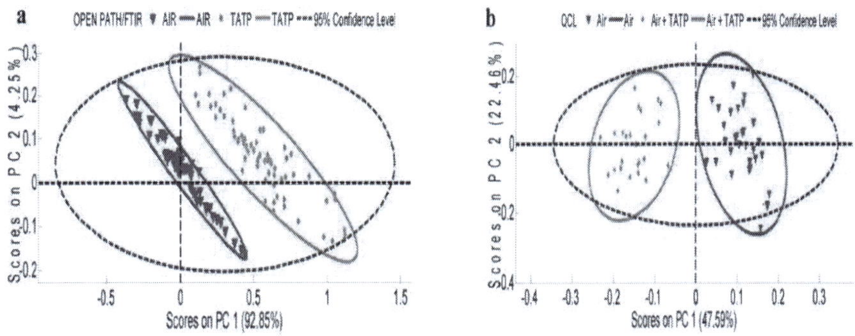

Figure 17. Score plots for the PCA, presented as (a) PC2 vs. PC1 for TATP detection from EM-27 FTIR spectrometer using Globar source and (b) PC2 vs. PC1 for TATP detection from LaserScope™ spectrometer using QCL source.

Other hand, the loadings plot were analyzed too to support the results from PCA with the finality of knowing which spectral signals cause differences between the dataset. Figure 18 shows the PC1 loading from Figure 17b, in this it can be seen than the spectral features are equal to infrared vibrational signal of reference TATP. Some signal recording can be tentatively assigned according to B. Brauer and J. Oxley as [47,48]: 891.8 cm^{-1} to O–C–O and Me–C–Me sym str, and Me–C–O asym str; 946 cm^{-1} to C-O str; 1197.6 cm^{-1} to O–C–O and Me–C–Me asym str, Me–C–O sym str; 1205 cm^{-1} to O–C–O and Me–C–Me sym str and finally 1234 cm^{-1} to C-C str,

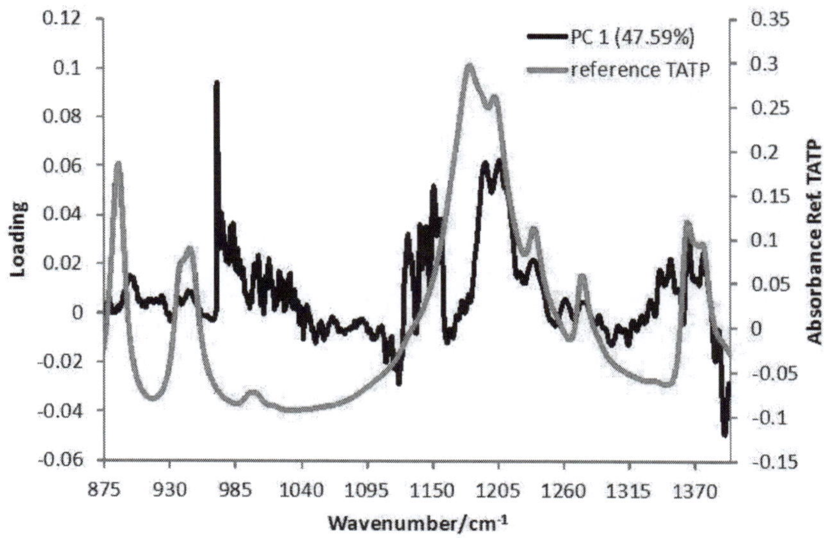

Figure 18. Loading plot for PC1 from PCA for TATP detection from LaserScope™ spectrometer using QCL source.

4.4. Quantum Cascade Laser Based Ir Reflectance Experiments

OPUS 6.0 Software (Bruker Optics, Billerica, MA, USA) was used to analyze the data obtained. Four spectra were obtained for each sample. The spectra were carried out using as backgrounds: substrate without explosive. PLS was applied to the data using different preprocessing treatments: raw data, auto scale. Mean center, auto scale + 1st derivative, auto scale + 2nd derivative, mean center + 1st derivative, mean center + 2nd derivative. The spectral range was 1000-1600 cm^{-1}. PLS shown below was that best results obtained. Figure 19 shows PLS plots of RDX deposited on TB. The best result was achieved using the spectral region of 1000-160 cm^{-1} and using mean centering as preprocessing. A total of 10 latent variables or factors were necessary to obtain a R^2 and RMSECV equal to 0.9915 and 2.32 µg/cm2, respectively.

Figure 20 shows PLS of PETN deposited on TB. The best resulted was achieved using the spectral region of 1000-1600 cm^{-1} and using mean centering as preprocessing treatment. A total of 10 latent variables or factors were necessary to obtain a R^2 and RMSECV equal to 0.9994 and 1.82 µg/cm2, respectively.

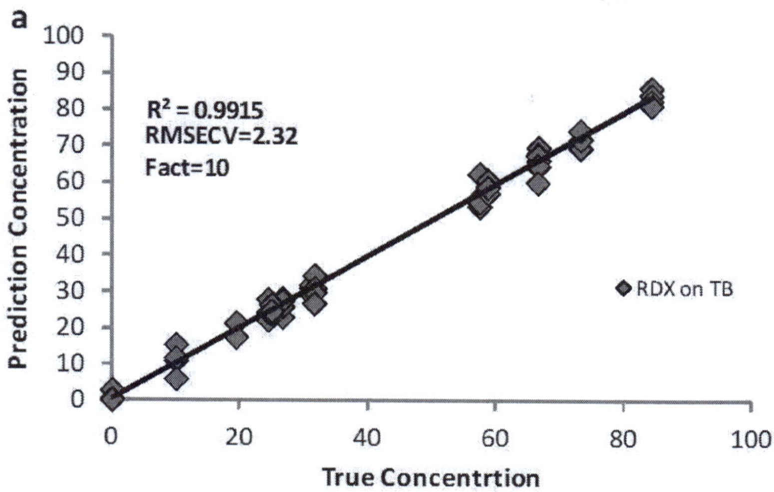

Figure 19. PLS of RDX on travel baggage (TB) as substrate.

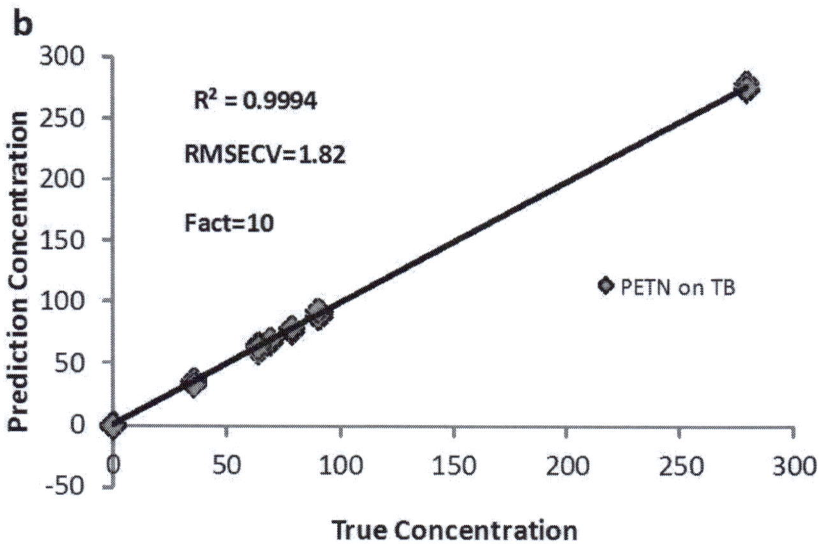

Figure 20. PLS of PETN on Travel Baggage as substrate.

4. CONCLUSION

Raman and infrared vibrational techniques were used for the detection of highly energetic materials and chemical warfare agents simulants in different matrices such as pharmaceutical mix, commercials bottles and travel baggage. The analysis of the spectral data allows emphasizing certain results. Satisfactory results were found for the quantification of explosives with good values of R^2cv, RMSECV. Reliable predictions obtained by remote sensing based on Raman spectroscopy at remote distance of 10 m employing 532 nm laser as excitation source. Remote Raman system using the appropriate chemometrics tools such as PLS, iPLS and siPLS promises to be a reliable technique for finding the existence of highly energetic material such as PETN deliberately hidden in matrices with similar chemical structures.

Partial Least Squares (PLS) calibration models reported limits of detection very low for white plastic in commercial beverage bottle solutions which was the best model. Due to the bottle material and commercial beverage product coloration, Malta was the worst model with reported limits of detection more elevated. Limits of Detection and Quantification for commercial bottles were compared in aqueous and mixtures. It is observed that limits of Detection were significantly lower for mixtures of TEP with the commercial product. Integration times were the same for both aqueous and commercial beverage bottle solutions (each normalized with respect to bottle material, color and thickness). Water does not transmit significant Raman signal, which would make limits of detection lower for aqueous solutions. However, commercial beverage bottles mixtures showed lower limits of detection than aqueous solution since the beverage solutions inside each bottle showed significant Raman signal and, therefore, increasing CWAS presence in the spectra.

PLS-DA model and discriminant analysis was done to detect TATP and 2,4-DNT traces in fluid air. The region of 600 to 7500 cm^{-1} was highly significant in the discrimination with $p < 0.00001$ and 100 % discrimination for two vectors for TATP and four vectors for 2,4-DNT. These results show the ability of the Chemometrics methods to discriminate between vapor phase explosive (2,4-DNT) and air.

Results obtained from principal component analysis to determine the presence of peroxides explosives such as TATP when they are in gas phase mixed with air shown be useful for distinction between TATP vapors and air. The principal component analysis from infrared spectral data used little PC for predict the variability of the spectral data, being the first two PC more important. PC1 loadings confirm the results from the PCA because it contained features from TATP spectrum. Other hand, the PLS model were shown chemometrics tool for quantify explosive such as RDX and PETN on substrate of the real world such as travel baggage.

In general, vibrational spectroscopy systems designed based on this work should be useful for National Defense and Security applications, for screening hazardous liquids in government installations, seaports and in public installations to improve defense against terrorist attacks.

ACKNOWLEDGEMENT

Support from the U.S. Department of Homeland Security under Award Number 2008-ST-061-ED0001 is also acknowledged. However, the views and conclusions contained in this document are those of the authors and should not be considered a representation of the official policies, either expressed or implied, of the U.S. Department of Homeland Security.

This contribution was supported by the U.S. Department of Defense, Proposal Number: 58949-PH-REP, Agreement Number: W911NF-11-1-0152. The authors also acknowledge contributions from Dr. Richard T. Hammond from Army Research Office, DOD.

REFERENCES

1. W. S Augerson, A Review of the Scientific Literature as it Pertains to Gulf War Illnesses 5Rand; 2000
2. A Vale, S Bradberry, P Rice, T. C Marrs, Chemical Warfare and Terrorism. The Medicine 20033192629
3. B Goozner, L. I Lutwick, E Bourke, Chemical Terrosism: a primer for 2002J. Assoc. Acad. Minor. Phys. 2002;1311418
4. M Rosenbloom, J. B Leikin, S. N Vogel, Z. A Chaudry, Biological and Chemical agents: A brief synopsis. Am. J. Ther. 200291514
5. C Gendering, Y Roggan, C Collet, Pharmaceutical applications of vibrational chemical imaging and chemometrics: A review. Journal of Pharmaceutical and Biomedical Analysis 2008483533553
6. M. L Ramirez, N Gaensbauer, H Felix, W Ortiz-rivera, L. C Pacheco-londono, S. P Hernandez-rivera, Fiber Optic Coupled Raman Based Detection of Hazardous Liquids Concealed in Commercial Products. International Journal of Spectroscopy 20121117
7. T Caron, M Guillemot, P Montméat, F Veignal, F Perraut, P Prené, F Serein-spirau, Ultra-trace detection of explosives in air: Development of a portable fluorescent detector. Talanta 2010

8. A Hilmi, J Luong, Micromachined Electrophoresis Chips with Electrochemical Detectors for Analysis of Explosive Compounds in Soil and Groundwater. Environ. Sci. Technol. 2000341430463050

9. J Yinon, Trace analysis of explosives in water by gas chromatography-mass spectrometry with a temperature-programmed injector. J. Chromatogr. A. 1996

10. C Szakal, T. M Brewer, Analysis and mechanisms of cyclotrimethylenetrinitramine ion formation in desorption electrospray ionization. Anal. Chem. 2009811352575266

11. C. J Miller, T. S Yoder, Explosive Contamination from Substrate Surfaces: Differences and Similarities in Contamination Techniques Using RDX and C-4. Sens. Imaging: An International Journal 20101127787

12. L. C Pacheco-londono, W Ortiz-rivera, O. M Primera-pedrozo, S. P Hernández-rivera, Vibrational spectroscopy standoff detection of explosives. Anal. Bioanal. Chem. 20093952323335

13. K Banas, A Banas, H. O Moser, M Bahou, W Li, P Yang, M Cholewa, S. K Lim, Multivariate Analysis Techniques in the Forensics Investigation of the Postblast Residues by Means of Fourier Transform-Infrared Spectroscopy. Anal. Chem. 201082730383044

14. C. W Van Neste, L. R Senesac, T Thundat, Standoff spectroscopy of surface adsorbed chemicals. Anal. Chem. 200981519521956

15. J Hildenbrand, J Herbst, J Wollenstein, A Lambrecht, M Razeghi, R Sudharsanan, G. J Brown, Explosive detection using infrared laser spectroscopy. Proc. SPIE 7222(1), 72220B; 2009

16. W Ortiz-rivera, L. C Pacheco-londono, S. P Hernandez-rivera, Remote Continuous Wave and Pulsed Laser Raman Detection of Chemical Warfare Agents Simulants and Toxic Industrial Compounds. Sensing and Imaging: An International Journal 2010113131145

17. W Ortiz, L. C Pacheco, J. R Castro, H Felix, S. P Hernandez-rivera, Vibrational spectroscopy standoff detection of threat chemicals. Proc. SPIE Int. Soc. Opt. Eng. 8031: 803129; 2011

18. J. R Castro, L. C Pacheco, M Vélez, M Diem, T. J Tague, S. P Hernandez, Open-Path FTIR Detection of Explosives on Metallic Surfaces. in Fourier Transforms: New Analytical Approaches and FTIR Strategies. G. S. Nikolić, ed. InTech Open, Croatia, 978-953-307-232-6; 2011

19. J. R Castro, L. C Pacheco, M Vélez, M Diem, T. J Tague, S. P Hernandez, Passive Mode FT-IR Standoff Detection of

Nitroaromatic High Explosives on Aluminum Substrates. (MS 11-06229R2Submitted and accepted in Applied Spectroscopy; 2012

20. O. M Primera, Y. M Soto, L. C Pacheco, S. P Hernández, Detection of High Explosives Using Reflection Absorption Infrared Spectroscopy with Fiber Coupled Grazing Angle Probe/FTIR. Sens. Imaging: An International Journal 2009

21. M. W Blades, H. G Schulze, S. O Konorov, C. J Addison, A. I Jirasek, R. F. B Turner, New Tools for Life Science Research Based on Fiber-Optic-Linked Raman and Resonance Raman Spectroscopy. ACS Symposium Series; 9632007

22. I. A Boere, Bakker Schut T.C., van den Booger J., de Bruin R.W.F., Puppels G.J. Use of fibre optic probes for detection of Barrett's epithelium in the rat oesophagus by Raman Spectroscopy. Vibrational Spectroscopy 20033214755

23. C Eliasson, N Macleod, P Matousek, Noninvasive Detection of Concealed Liquid Explosives Using Raman Spectroscopy. Analytical Chemistry 2007792181858189

24. R. F Curl, F Capasso, C Gmachl, A. A Kosterev, B Mcmanus, R Lewicki, M Pusharsky, G Wysocki, F. K Tittel, Quantum cascade lasers in chemical physics. Chem. Phys. Lett. 2010

25. J Hildenbrand, J Herbst, J Wollenstein, A Lambrecht, M Razeghi, R Sudharsanan, G. J Brown, Explosive detection using infrared laser spectroscopy. Proc. SPIE 7222(1), 72220B; 2009

26. B Hinkov, F Fuchs, J. M Kaster, Q Yang, W Bronner, R Aidam, K Kohler, J. C Carrano, C. J Collins, Broad band tunable quantum cascade lasers for stand-off detection of explosives. Proc. SPIE 7484(1), 748406; 2009

27. T Urbanski, Chemistry and technology of explosives. New York: Macmillan Company; 1964

28. K. R Beebe, R. J Pell, M. B Seasholtz, Chemometrics: A Pactrical Guide. New York: John Wiley & Sons; 1998

29. R. G Brereton, Applied Chemometrics for Scientists. The Atrium, Southern Gate: John Wiley & Sons Ltd. Chichester; 2007

30. D. L Massart, B. G. M Vandeginste, L. M. C Buydens, Data Handling in Science and Technology Handbook of Chemometrics and Qualimetrics Part B. The Netherlands: Elsevier Science B.1997

31. D. L Massart, B. G. M Vandeginste, L. M. C Buydens, Handling in Science and Technology Handbook of Chemometrics and Qualimetrics Part A. The Netherlands: Elsevier Science, B.1997

32. L Norgaard, A Saudland, J Wagner, J. P Nielsen, L Munck, S. B Engelsen, Interval Partial Least Squares Regression (iPLS): A Comparative Chemometrics Study with an Example from Near-Infrared Spectroscopy, Applied Spectroscopy 2000543413419

33. R Leardi, L Norgaard, Sequential application of backward interval PLS and Genetic Algorithms for the selection of relevant spectral regions. Journal of Chemometrics 20041811486497

34. I Delfino, C Camerlingo, M Portaccio, Della Ventura B., Mita L., Mita D.G., Lepore M. Visible micro-Raman spectroscopy for determining glucose content in beverage industry. Food Chemistry 20111272735742

35. L Norgaard, M Hahn, L. B Knudsen, I. A Farhat, S. B Engelsen, Multivariate near-infrared and Raman spectroscopic quantifications of the crystallinity of lactose in whey permeate powder. International Dairy Journal 2005151212611270

36. Q Chena, J Zhao, M Liu, C Jianrong, L Jianhua, Determination of total polyphenols content in green tea using FT-NIR spectroscopy and different PLS algorithms. Journal of Pharmaceutical and Biomedical Analysis 2008463568573

37. Y. A Gruzdkov, Y. M. J Gupta, Vibrational Properties and Structure of Pentaerythritol Tetranitrate. Phys. Chem. A. 20011052561976202

38. I. R Lewis, N. W Daniel, P. R Griffits, Interpretation of Raman Spectra of Nitro Containing Explosive Materials. Part I: Group Frequency and Structural Class Membership. Applied Spectroscopy 1997511854

39. J. P Pestaner, F. G Mullick, J. A Centeno, Characterization of acetaminophen: molecular microanalysis with Raman microprobe spectroscopy. J. Forensic. Sci. 1996411060

40. M Felipe, M. J Cal, J Ferre, R Boque, J. M Andrade, A Carlosena, Linear PLS regression to cope with interferences of major concomitants in the determination of antimony by ETAAS. J. Anal. At. Spectrom. 20062116168

41. J. N Miller, J. C Miller, Statistics and Chemometrics for Analytical Chemistry. Fourth edition, Pearson Prentice Hall; 2000

42. G. J Mclachlan, Discriminant Analysis and Statistical Pattern Recognition. Hoboken, New Jersey: John Wiley & Sons Inc. Chichester; 1992

43. C. L Liu, H Sako, H Fujisawa, Discriminative Learning Quadratic Discriminant Function for Handwriting Recognition. IEEE Transactions on Neural Networks 2004

44. B. K Alsberg, D. B Kell, R Goodacre, Variable Selection in Discriminant Partial Least-Squares Analysis. Anal. Chem. 1998701941264133

45. R. G Brereton, Chemometrics. Data Analysis for Laboratory and Chemical Plant. The Atrium, Southern Gate: John Wiley & Sons Ltd. Chichester; 2003

46. J Olivero, R Vivas, L Pacheco, B Johnson, K Kannan, Discriminant analysis for activation of the aryl hydrocarbon receptor by polychlorinated naphthalenes. Journal of Molecular Structure: Theochem. 200467813157161

47. B Brauera, F Dubnikova, Y Zeiri, R Kosloff, R. B Gerbera, Vibrational spectroscopy of triacetone triperoxide (TATP): Anharmonic fundamentals, overtones and combination bands. Spectrochimica Acta Part A. 2008711438

48. J Oxley, J Smith, J Brady, F Dubnikova, R Kosloff, L Zeiri, Y Zeiri, Raman and Infrared Fingerprint Spectroscopy of Peroxide-Based Explosives. Applied Spectroscopy 2008628906915

8

CHAPTER

Evaluation of Major Factors Influencing the Geochemistry of Groundwater Using Graphical and Multivariate Statistical Methods in Yenagoa City, Southern Nigeria

K. S. Okiongbo [1] • R. K. Douglas[2]

[1] Department of Geology and Physics, Niger Delta University, Wilberforce Island, Bayelsa State, Nigeria

[2] School of Applied Sciences, Cranfield University, Cranfield, Bedforshire MK 43 0AL, UK

ABSTRACT

To achieve a better understanding of the nature of the factors influencing groundwater composition as well as to specify them quantitatively, conventional graphical and multivariate statistical analysis (principal component analysis) were applied on hydrochemical data consisting of 51 groundwater samples collected from domestic boreholes in Yenagoa city, Bayelsa State, Nigeria. The mode of study includes analysis of major ion contents and other chemical parameters such as pH, total dissolved solids and electrical conductivity of the groundwater samples. The PCA yielded three principal components explaining 78.38 % of the total variance of the 11 parameters. The three components are interpreted as controlled by the natural weathering of existing silicate rocks, reverse ion-exchange processes and oxidation reactions which are further supported by the scatter diagrams, ionic signatures and mechanisms controlling the water chemistry diagrams as the common factors influencing the groundwater hydrogeochemical character. Limited anthropogenic influence on the groundwater composition has also been noticed in the study area. The groundwater poses no threat to human health because the concentrations of physico-chemical parameters that can be used to evaluate drinking water quality are within World Health Organisation standard specification. The groundwater in the area is fresh, high salinity and low sodium in nature.

Keywords: Multivariate statistical analysis R-mode factor analysis Groundwater quality Hydrogeochemistry Yenagoa

INTRODUCTION

Hydrogeochemical investigations to characterise groundwater and assess its chemical quality often involve substantially large amount of base information concerning groundwater chemistry. Groundwater chemistry, in turn, depends on a large number of factors, such as general geology, degree of chemical weathering of the various rock types, quality of recharge water and inputs from sources other than water–rock interaction. Such factors and their interactions result in a complex groundwater quality. Thus, the characterisation and interpretation of various groundwater parameters is often a complex problem. Several conventional methods of data analysis are often used for the interpretation and presentation of results (Hem1989), e.g. histograms, stiff, trilinear, semilogarithmic diagrams and many others (Lloyd and Heathcote 1985). The use of these graphical methods to interpret aquatic chemistry is limited to two dimensions, besides, deal only with a limited number of variables responsible for the groundwater chemistry and can produce biased results (Guler et al. 2002).

To overcome the limitations of these conventional methods, multivariate statistical analysis such as principal component analysis had been widely used for characterizing and interpreting geochemical data. Such analyses are especially useful because in addition to revealing the interaction among the variables, it also attempts to simplify the complex and diverse relationships that exist among a set of observed variables (i.e. the concentrations of physico-chemical constituents) by revealing common and unobserved factors (i.e. the underlying physico-chemical constituents and/or biological processes in the groundwater system) that link together with the seemingly unrelated variables (Lee et al. 2001). Reeder et al. (1972) identified the likely weathering processes controlling the chemical composition of surface waters of Mackenzie River drainage basin Canada. Ashley and Lloyd (1978) used factor analysis to evaluate the hydrogeochemical process in the Santiago basin of Chile and the Derbyshire Dome of England. Lawrence and Upchurch (1982) also used the technique to delineate the zones of natural recharge to groundwater in the Florida aquifer. Ruiz et al. (1990) suggested that the basic purpose of such an analysis was to study the hydrogeochemistry of aquifers by simplifying the numerous and complex groundwater data into a set of factors, few in number which can explain a large amount of the variance of the analytical data and also indicate the source of origin of various ions present in water. Earlier studies carried out in the study area are on the geochemistry of groundwater (Amadi et al. 1987), prospecting of groundwater resources (Okiongbo and Ogobiri 2011), and groundwater quality with respect to drinking and agricultural purposes (Okiongbo and Douglas 2013). These studies mainly focus on chemical composition of groundwater using conventional methods only.

In this study, we carry out a hydrochemical evaluation of an alluvial aquifer system with an objective of delineating the various factors controlling the water chemistry and general suitability of the groundwater for domestic and drinking purposes using conventional graphical and multivariate statistical methods.

DESCRIPTION OF THE STUDY AREA

Location and climate

The study area entails Yenagoa city, part of Southern Ijaw and Ogbia Local Government Areas in Bayelsa State and lies between longitudes 006°05′ and 00 6°040′ East of the prime meridian and latitudes 04°23.3′ and 04°38.2′ North of the equator within the coastal area of the recent Niger Delta (Fig. 1). Bayelsa State is located in the southern end of Nigeria and is within the Niger Delta Basin. It is bounded in the north by Rivers State, in the west by Delta State and in the south by the Bight of Bonny. The study area has an areal extent of about 150 km^2. The study area has a tropical rain forest climate characterised by two seasons, namely the wet or rainy season and the dry season. The rainy season lasts for about 7 months between April and October with an intervening dry period in August. The dry season lasts for about 4 months, between November and March. The temperature varies between 25 and 32 °C. The mean annual rainfall is about 4,500 mm (Akpokodje 1986); about 85 % of the mean annual rain falls in the wet season.

PHYSIOGRAPHY AND DRAINAGE

The study area consists of alluvial deposits and an extensive, low-lying, typical deltaic plain with essentially flat topography which, in conjunction with the high annual rainfall, is responsible for the extremely poor drainage conditions and the widespread development of marshes and backswamps. This area is usually submerged during the wet season where flood waters range from 0.5 to 4 m deep (Akpokodje 1986). There are a number of perennial streams, oxbow lakes and rivers in the area e.g. Kolo Creek, Epie

Creek, Yenagoa and Nun river, etc. They all form a network which empties to the Atlantic Ocean through Nun River Estuary. These rivers are mostly turbid during the wet season possibly due to discharge of clay and silt (Amadi et al.1987). The natural vegetation of the study area is that of the rain forest but this has been destroyed by the activities of man such as bush burning, farming, construction and illegal crude oil refining activities. The vegetation consists of various kinds of evergreen trees, including palms trees and a variety of shrubs. More than 70 % of the inhabitants of the study area are engaged in subsistent farming and fishing.

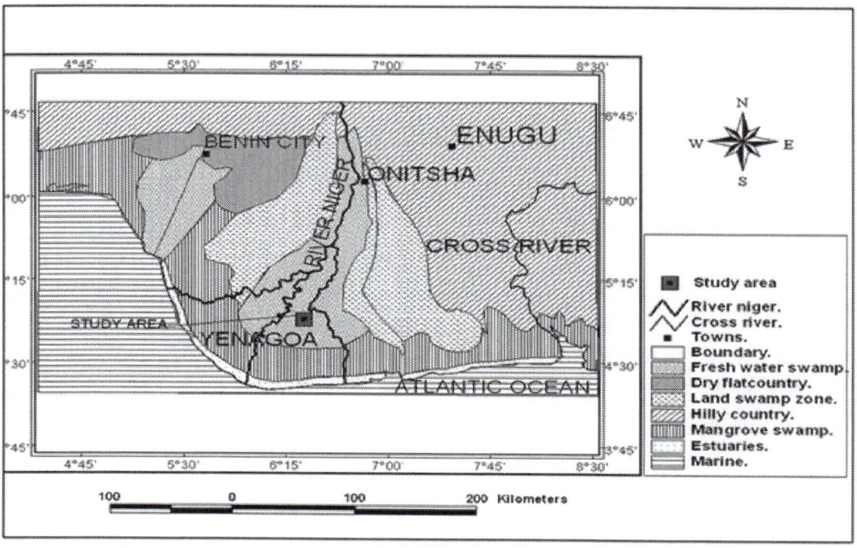

Figure 1. The major geomorphological units of the Niger Delta (After Akpokodje 1986)

GEOLOGY

The study area (Fig. 1) lies within the fresh water forests, swamps and backswamp geomorphic unit of the Niger Delta. The Niger Delta is basically an alluvial plain formed by the deposition and built-up of fine-grained sediments eroded and transported to the area by the River Niger and its tributaries (Fig. 1). The regional geology of the Niger Delta is relatively simple consisting of three lithostratigraphic units, Akata, Agbada and Benin Formations, overlain by various types of quaternary deposits (Short and Stauble 1967; Wright et al. 1985; Kogbe 1989).

The quaternary deposits are considered universally to be the recent expressions of and continuation of the Benin Formation and consist of medium to coarse-grained sands, sandy clays, silts and subordinate, lensoid clay bands thought to have been deposited during quaternary interglacial marine transgressions (Durotoye 1989). Amajor (1991) has shown that they are an admixture of fluvial/tidal channel, tidal flats and mangrove swamp deposits. The sands are micaceous and feldspathic, subrounded to angular in texture and constitute good aquifers. However, depth to occurrence and thickness is irregular and may not be predicted with accuracy within the study area due to rapid horizontal and vertical facie changes. Groundwater occurs generally under water table conditions where the lensoid clays attain adequate thickness.

The Benin Formation is essentially fluvial in origin comprising unconsolidated, massive and porous fresh water bearing sands with localized clay/shale beds. The sand-clay intercalations are indicative of a multi-aquifer system. The thickness of the Benin Formation is about 2,100 m on the average. The age of this Formation ranges from Miocene to Recent. Onyeagocha (1980) reported that the rocks are made up of 95 % quartz grains, Na + mica 1–2.5 %, feldspar 0–1.0 % and dark coloured minerals 2.3 %. Water table in large sections of the Niger Delta is close to the surface but subject to spatial and seasonal variations. In the study area, the water table is about 3–4 m during the dry season. During the wet season, the water table rises considerably, in some cases, to the ground surface. The Agbada Formation underlies the Benin Formation. Its thickness varies from 300 to 4,500 m (Short and Stauble1967). This Formation consists predominantly of unconsolidated pebbles, and very coarse to fine-grained sand units with subordinate shale beds. Underneath the Agbada Formation is the Akata Formation of Eocene to recent age (Amadi et al. 1987), which is made up of shales with local inter-bedding of sands and siltstones. It was deposited in a typically marine environment with maximum thickness of about 6,100 m (Ofoegbu1989).

HYDROGEOLOGY

The Coastal Plain Sands of the Benin Formation are the main regional and most important aquifer in the study area. Groundwater in the Coastal Plain Sands occurs mainly under phreatic (unconfined) conditions. The lithology of this formation is dominated by loose sands (fine-medium-coarse), while gravel and pebbles are minor components. Thin clay horizons and lenses create discontinuities in the vertical and lateral continuity of the porous sands and gravel. This condition results in the presence of local perched aquifers. Rainfall is the direct recharge source of the groundwater. The transmissivity values for the aquifer range from 1.05×10^{-2} to $11.3 \times 10^{-2} m^2/s$, while the coefficient of storage varies between 1.07×10^{-4} and 3.53×10^{-4} and specific capacity values lie between 19.01 and 139.8 $m^3/h/m$ drawdown (Amajor 1991). These values suggest that the aquifers have very good capacity to transmit groundwater. The groundwater quality is very good and compares favourably with WHO standards for drinking water. However, relatively high iron and chloride values though localized are observed.

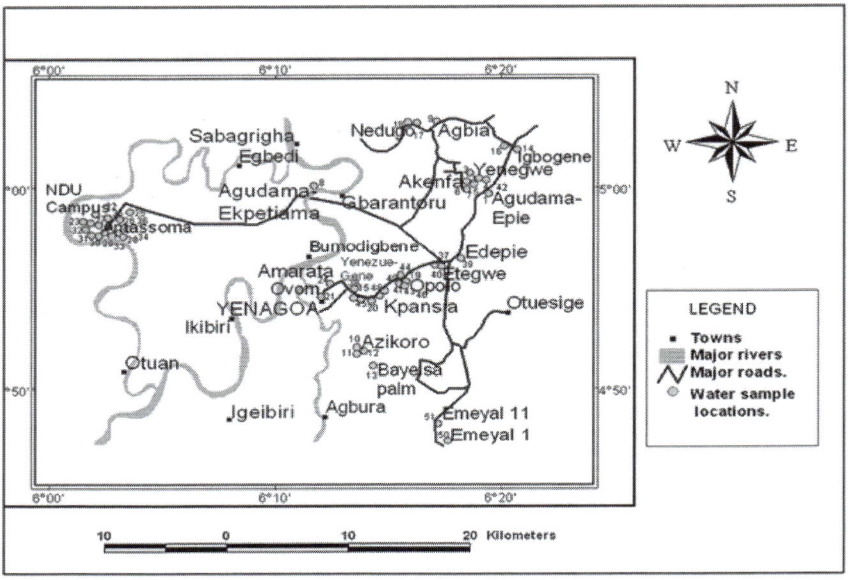

Figure 2. Map of study area showing borehole locations

METHOD OF STUDY

Groundwater samples were collected from 51 domestic boreholes spread over the study area (Fig. 2) in 3 weeks in the month of February, 2013, and were analysed within 24 h. A global positioning system (GPS) was used for location and elevation readings. The samples were collected in polyethylene bottles for the analysis of total hardness, total dissolved solids (TDS) and major ions such as Na, K, Ca, Mg, HCO_3, CO_3, Cl and SO_4 in the laboratory. All samples were analysed using standard procedures (APHA 1995). Electrical conductivity (EC) and pH of groundwater samples were measured in the field immediately after sample collection using portable field kits (Hach's CO 150 Conductivity/TDS metre and pH metre, respectively). Major ions like sulphate (SO_4^{2-}) were determined by spectrophotometric turbidimetry. The method is based on the principle of formation of barium sulphate in the presence of (acidified HCl) barium chloride. The process is enhanced in the presence of glycerol or other organic compound. The absorbance of the colloidal solution was measured against a standard on

UV-visible Spectrophotometer. Calcium (Ca^{2+}) and magnesium (Mg^{2+}) were determined titrimetrically using EDTA; chloride (Cl^{-}) was determined by standard $AgNO_3$ titration; bicarbonate (HCO_3^{-}) was determined by titration with HCL; sodium (Na^{+}) and potassium (K^{+}) were measured by flame photometry; nitrate (NO_3^{-}) was determined by colourimetry with a UV-visible spectrophotometer (brucine method) while iron was determined by colourimetry with a UV-visible spectrophotometer at 520 nm. Iron results represent ferrous iron. In general, ion balance error of the samples was observed to be within the acceptable range of ±10 %.

The data was analysed using multivariate statistical technique (principal component analysis) to identify the hydrogeochemical processes which result in the variations in the chemical composition of groundwater. Saturation indices of groundwater with respect to the mineral phases and the chemical facies were determined using the computer programmes PHREEQC and AQUACHEM, respectively.

The obtained matrix of hydrogeochemical data was subjected to multivariate analytical technique. Principal component analysis (PCA) was used here as a numerical method of discovering variables that are more important than others for representing parameter variation and identifying hydrogeochemical processes. The entire dataset was first standardized and arranged in correlation coefficient matrix with normal distribution in all variables. The principal component approach started by extracting eigenvalues and eigenvectors of the correlation matrix and then less important of these was discarded (Davis 1986). The eigenvalues calculated quantify the contribution of a factor to the total variance. The percentages were also computed. The contribution of a factor is significant when the eigenvalue is greater than unity. The factors that are significant (with eigenvalues more than one and explaining for the higher percentage of the variability of the data) were noted. The factor loadings were calculated by a varimax rotation technique in such a way that they are closer to +1, 0, −1, representing positive contribution, no contribution and negative contribution. They suggest the dominance of the corresponding variables which, when projected onto the data, give factor scores.

RESULTS AND DISCUSSION

Groundwater chemistry

The physico-chemical compositions of the groundwater samples are given in Table 1, and Table 2 gives the descriptive statistical measures such as minimum, maximum, mean and standard deviation values of various parameters investigated along with the World Health Organisation standards (WHO 2004). In the groundwater samples, pH value ranges from

Table 1. Hydrogeochemical analysis of groundwater samples

S/N	Code	pH	EC	TDS	TA	TH	Na^+	K^+	Ca^{2+}	Mg^{2+}	Fe	Cl^-	SO_4^{2-}	NO_3^-	HCO_3	CAI-1	CAI-2	SO4/Cl	SAR
1	BH1	6.83	2,120.0	1,060.0	55.0	143.0	52.2	30.4	40.5	10.7	0.50	31.8	7.2	0.13	300.8	−2.4	−0.4	0.23	2.7
2	BH2	6.16	338.0	169.0	15.0	48.0	14.2	6.8	2.4	0.5	0.02	2.9	0.5	0.11	52.5	−8.9	−0.8	0.18	3.1
3	BH3	5.78	615.0	308.0	26.0	25.0	28.4	16.5	4.0	0.8	0.01	4.8	4.2	0.08	103.7	−11.2	−0.9	0.87	4.8
4	BH4	5.87	472.0	236.0	16.0	16.0	14.5	12.0	6.1	3.5	0.18	4.6	1.2	0.08	84.2	−6.2	−0.6	0.26	1.6
5	BH5	6.17	400.0	200.0	20.0	6.0	18.2	9.9	5.4	3.1	0.15	12.5	1.8	0.04	72.0	−2.0	−0.6	0.14	2.2
6	BH6	5.82	768.0	384.0	39.0	50.0	16.5	6.8	5.5	2.5	0.24	5.5	4.6	0.27	68.3	−4.7	−0.6	0.83	2.1
7	BH7	6.03	515.0	256.0	21.0	44.0	18.4	7.5	3.5	1.2	0.18	4.8	1.7	0.16	67.1	−6.3	−0.8	0.35	3.1
8	BH8	6.38	708.0	354.0	30.0	48.0	28.4	9.5	6.8	3.4	0.12	17.3	2.8	0.03	94.6	−2.0	−0.6	0.16	3.1
9	BH9	6.20	2,620.0	1,310.0	48.0	254.0	60.4	28.5	36.4	8.5	0.02	82.0	5.7	0.13	209.8	−0.5	−0.3	0.07	3.3
10	BH10	6.24	800.0	400.0	28.0	40.0	14.5	8.7	5.6	3.5	0.25	11.3	0.9	0.16	66.5	−1.7	−0.5	0.08	1.7
11	BH11	6.15	1,036.0	518.0	24.0	40.0	29.5	18.4	6.3	2.7	0.05	14.4	3.2	0.14	111.0	−3.3	−0.7	0.22	3.5
12	BH12	5.92	907.0	454.0	42.0	57.0	25.5	15.6	8.5	2.2	0.29	14.8	4.2	0.05	98.2	−2.6	−0.6	0.28	2.9
13	BH13	6.16	696.0	348.0	42.0	52.0	12.5	8.2	4.5	5.6	0.32	14.5	2.5	0.01	60.4	−0.8	−0.3	0.17	1.3
14	BH14	5.80	1,091.0	546.0	33.0	120.0	42.2	17.6	5.5	2.3	0.18	21.2	4.0	0.09	126.9	−2.8	−0.8	0.19	5.4
15	BH15	6.31	418.0	209.0	30.0	15.0	12.2	10.5	6.74	3.46	0.25	16.0	0.54	0.09	59.2	−0.8	−0.4	0.03	1.3
16	BH16	5.87	1,324.0	662.0	60.0	54.0	41.7	18.4	12.8	8.8	0.21	20.4	3.5	0.14	183.0	−3.0	−0.6	0.17	3.1
17	BH17	5.87	1,098.0	549.0	40.0	48.0	15.4	9.8	10.7	6.2	0.31	21.6	10.6	0.00	70.2	−0.5	−0.2	0.49	1.3
18	BH18	6.00	440.0	220.0	39.0	23.5	15.2	7.4	4.6	2.5	0.38	11.7	5.5	0.28	51.9	−1.6	−0.5	0.47	2.0
19	BH19	6.04	207.0	104.0	13.0	14.2	3.8	2.5	2.9	1.6	0.32	10.3	2.1	0.20	11.0	0.2	0.3	0.20	0.6
20	BH20	6.34	2,820.0	1,410.0	65.0	265.0	48.7	15.2	20.9	6.9	0.18	39.1	6.5	0.42	175.7	−1.3	−0.5	0.17	3.3
21	BH21	6.88	792.0	396.0	30.0	68.0	18.8	8.2	8.0	1.8	0.18	8.4	5.4	0.15	70.0	−3.4	−0.6	0.65	2.2
22	BH22	6.18	495.0	248.0	18.0	42.0	13.8	6.5	3.5	1.8	0.86	14.2	4.7	1.60	36.0	−0.9	−0.5	0.33	2.1
23	BH23	6.19	409.0	205.0	32.0	23.0	17.2	8.3	5.0	4.5	0.28	6.9	2.6	0.05	81.7	−3.9	−0.5	0.38	1.9
24	BH24	6.70	591.0	296.0	26.0	24.0	7.8	12.0	5.7	4.2	0.20	5.3	3.3	0.28	64.7	−3.3	−0.4	0.62	0.9
25	BH25	6.06	910.00	455.00	23.00	60.00	12.40	7.60	13.50	3.80	10.10	12.50	2.30	6.74	8.40	−1.08	−1.15	0.18	1.1
26	BH26	5.80	890.00	450.00	17.00	85.00	10.60	5.80	14.80	12.60	2.08	28.40	1.20	1.68	6.20	0.24	1.17	0.04	0.7
27	BH27	6.20	1,060.00	530.00	22.00	57.00	12.50	10.50	28.50	21.80	11.94	49.70	3.00	5.85	10.80	0.42	1.60	0.06	0.6
28	BH28	6.22	1,200.00	600.00	26.00	75.00	9.50	12.60	18.80	6.70	12.00	17.70	5.40	8.56	18.20	−0.47	−0.40	0.31	0.7
29	BH29	6.19	690.00	345.00	16.00	55.00	8.50	4.60	15.70	6.50	10.61	24.50	1.10	1.50	5.60	0.29	1.38	0.04	0.6
30	BH30	6.44	1,163.00	582.00	26.00	65.00	12.30	6.40	16.50	10.80	6.06	21.30	6.50	2.45	14.80	−0.16	−0.23	0.31	0.8
31	BH31	6.46	1,089.00	545.00	15.00	75.00	8.80	9.20	15.20	8.00	7.38	14.20	4.80	3.40	9.40	−0.54	−0.66	0.34	0.6
32	BH32	6.65	1,138.00	569.00	18.00	166.00	10.50	9.60	8.80	6.50	6.24	12.40	3.50	1.30	8.80	−1.01	−1.44	0.28	0.9
33	BH33	6.04	526.00	520.00	25.00	4.00	8.40	9.60	6.80	7.20	12.00	10.50	0.50	2.70	12.70	−1.06	−1.14	0.05	0.8
34	BH34	6.38	1,068.00	534.00	15.00	73.00	7.40	7.50	10.20	5.80	12.08	17.00	0.84	4.85	10.50	−0.07	−0.12	0.05	0.6
35	BH35	6.84	591.00	296.00	38.00	37.00	6.50	7.80	10.80	8.20	9.08	30.40	0.38	2.20	4.00	0.44	3.08	0.01	0.5
36	BH36	6.93	593.00	297.00	10.00	15.00	6.40	7.40	2.80	6.90	12.10	15.00	0.54	3.70	3.80	−0.11	−0.29	0.04	0.7
37	BH37	6.31	2,250.00	1,125.00	42.00	260.00	38.90	14.00	15.00	5.70	12.26	18.50	12.40	4.60	40.00	−2.93	−1.51	0.67	3.1
38	BH38	6.02	277.00	139.00	12.40	13.50	15.60	10.50	10.20	6.40	1.20	21.00	8.40	0.32	4.80	−0.60	−1.36	0.40	1.3
39	BH39	5.72	191.00	96.00	8.60	6.80	8.50	4.60	5.40	3.20	3.40	9.80	3.80	0.12	2.80	−0.76	−1.65	0.39	1.0
40	BH40	5.77	508.00	254.00	14.50	15.00	17.70	14.20	12.60	6.80	0.15	19.90	15.40	0.35	8.60	−1.02	−1.22	0.77	1.4
41	BH41	6.00	331.00	165.00	15.00	19.00	8.50	6.70	11.40	7.60	0.22	22.80	14.60	0.16	6.20	0.16	0.25	0.64	0.7
42	BH42	5.78	201.00	100.00	13.50	16.70	7.50	5.40	8.60	4.80	4.60	18.20	13.60	0.15	5.80	0.09	0.13	0.75	0.7
43	BH43	5.70	323.00	162.00	14.20	9.80	26.50	12.70	6.25	3.24	5.20	22.20	16.50	0.13	6.90	−1.36	−1.85	0.74	3.0
44	BH44	5.69	348.00	174.00	15.10	14.00	18.30	10.88	8.94	6.41	2.50	20.00	15.40	0.29	7.20	−0.91	−1.15	0.77	1.6
45	BH45	5.71	335.00	168.00	12.00	11.70	7.80	4.10	7.52	4.70	3.40	17.70	15.30	0.20	6.70	0.11	0.13	0.86	0.8
46	BH46	5.64	119.00	60.00	2.50	3.50	3.80	3.50	3.40	2.85	4.80	16.50	9.54	0.12	3.80	0.45	0.80	0.58	0.5
47	BH47	5.86	161.00	82.00	3.80	4.70	1.07	4.70	4.85	3.42	2.60	18.50	9.35	0.14	2.80	0.68	1.45	0.51	0.1
48	BH48	6.16	102.00	51.00	2.00	7.50	1.42	1.02	1.20	0.85	0.38	19.10	0.85	0.15	1.60	0.84	9.54	0.04	0.3
49	BH49	5.55	294.00	147.00	13.00	12.60	4.28	1.85	3.50	2.80	0.25	12.30	12.60	0.28	5.50	0.33	0.32	1.02	0.8
50	BH50	6.35	112.00	60.00	2.20	3.00	2.80	2.00	2.40	1.60	0.42	36.50	16.72	0.07	2.00	0.83	2.23	0.46	0.5
51	BH51	6.33	909.00	455.00	28.50	18.60	38.50	12.60	18.50	6.70	3.50	54.80	28.80	0.28	14.50	−0.29	−0.54	0.53	2.8

All parameters have been expressed as mg/L, except pH and EC. The unit of EC is µS/cm while total alkalinity (TA) and total hardness (TH) is mg/L $CaCO_3$. The iron results represent ferrous iron and SAR is expressed in meq/l

5.55 to 6.93 with an average of 6.93. The decrease in pH is primarily due to oxidation of organic matter and also due to anthropogenic activities (Krishnaraj et al. 2011). Concentrations of TDS, a measure of groundwater quality ranged from 51 to 1,410 mg/L with an average of 388.3 mg/L. According to TDS classification, 90 % of the samples belong to fresh water type (TDS <1,000 mg/L) and the remaining 10 % represent brackish water (TDS > 1,000 mg/L). The electrical conductivity values range from 102 to 2,820 µS/cm, with an average of 765.9 µS/cm.

Table 2. Descriptive statistics of analysed groundwater samples compared with standards

Parameter	No of Samples	Minimum	Maximum	Mean	Standard Deviation	WHO 2004 maximum permissible limit
pH	51	5.55	6.93	6.13	0.34	6.5–8.5
EC (μS/cm)	51	102	2,820	765.9	601.78	1,400
TDS (mg/L)	51	51	1,410	388.3	300.87	1,000
TA (mg/L)	51	2	65	24.4	14.46	
TH (mg/L)	51	3	265	53	62.60	500
Na^+ (mg/L)	51	1.1	60.4	17.4	13.65	200
K^+ (mg/L)	51	1.02	30.4	9.9	5.83	55
Ca^{2+} (mg/L)	51	1.2	40.5	9.9	8.05	75
Mg^{2+} (mg/L)	51	0.5	21.8	5.2	3.66	50
Fe (mg/L)	51	0.01	12.3	3.18	4.27	0.3
Cl^- (mg/L)	51	2.9	54.8	19.2	13.80	250
SO_4^{2-} (mg/L)	51	0.5	28.8	6.1	5.89	400
NO_3^- (mg/L)	51	0.01	8.56	1.12	1.94	50
HCO_3^- (mg/L)	51	6.2	300.8	50	62.13	–

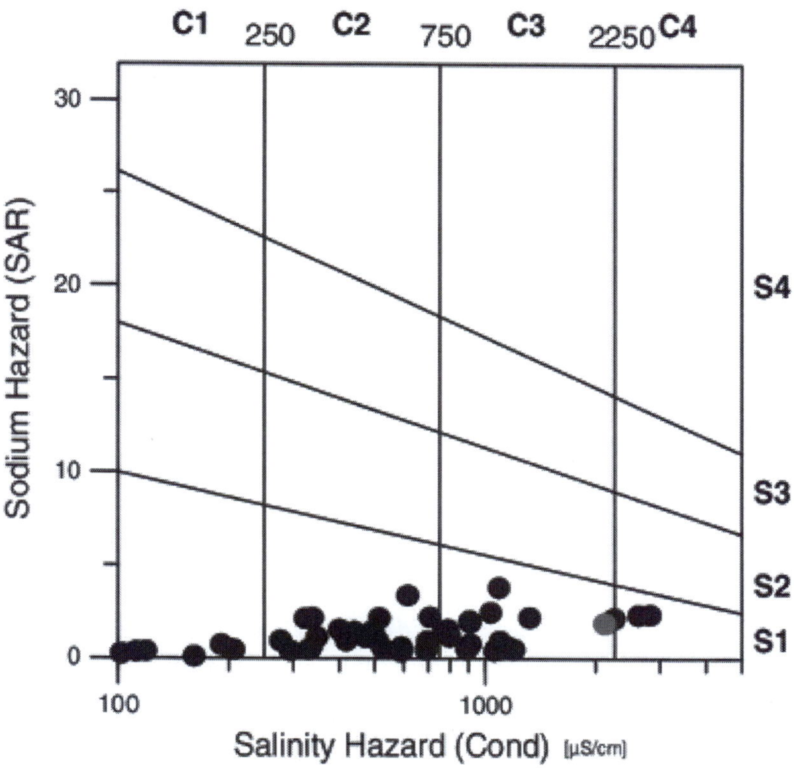

Figure 3. Wilcox salinity diagram

Tables 1 and 2 show that the major cations Na^+, K^+, Ca^{2+}, Mg^{2+} and Fe are all low except iron with mean concentrations of 17.4, 9.9, 9.9, 5.2 and

3.18 mg/L, respectively. The major anions HCO_3^-, Cl^-, $SO_4^{2-}$and NO_3^- are also low with mean concentrations of 50 mg/L, 19.2 mg/L, 6.1 mg/L and 1.12 mg/L, respectively. The results obtained show that the concentration of dissolved major cations and anions in the groundwater varies spatially and mass abundance was in the order of: $Na^+ > Ca^{2+} > K^+ > Mg^{2+} > Fe$ for cations and $HCO_3^- > Cl^- > SO_4^{2-} > NO_3^-$ for anions. The average concentrations of the major ions in the groundwater indicate that the groundwater of the area is suitable for drinking in all samples except iron. The analytical data plotted on the EC versus SAR (sodium adsorption ratio = $Na^+/([Ca^{2+} + Mg^{2+}]/2)^{0.5}$ diagram (Fig. 3) illustrates that most of the groundwater samples fall mainly in the field of C2–S1 and C3–S1 indicating medium to high salinity and low sodium water. Thus, the groundwater in the area can be used for irrigation on almost all types of soil with little danger of exchangeable sodium.

A Piper diagram was plotted to examine the hydrochemical water types (Fig. 4). The Piper diagram shows that the chemical composition of the groundwater in the study area is composed of six main types: Ca–Cl, Ca–HCO_3–Cl, Ca–Na-Cl, Ca–Na–HCO_3–Cl, Ca–Na–HCO_3 and Na-HCO_3. About 52.2 % groundwater is dominated by the alkalis (Na^+ and K^+) over the alkaline earths (Ca^{2+} and Mg^{2+}) and in 54.2 % of samples, weak acids (HCO_3^-) exceed strong acids (SO_4^- and Cl^-).

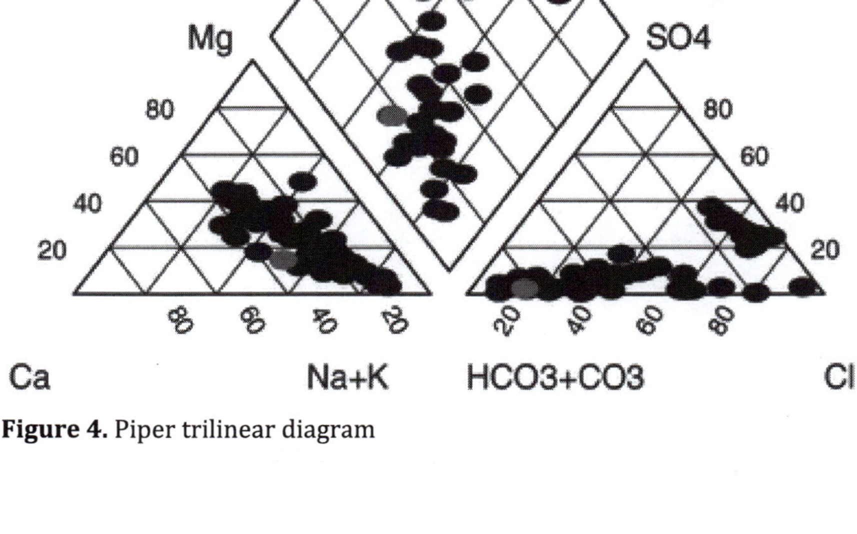

Figure 4. Piper trilinear diagram

Hydrogeochemical evaluation

The geochemical variations in the ionic concentrations in the groundwater can easily be understood when they are plotted along an *X–Y* coordinate (Guler et al. 2002). Results from the chemical analyses were used to identify the geochemical processes and mechanisms in the groundwater aquifer system. Gibbs plots of TDS versus the ion ratios Na/(Na + Ca) and Cl/(Cl + HCO_3) (Fig. 5) show that all data points fall on the intermediate area between the rock dominance end member and evaporation/precipitation dominance end member indicating that rock weathering was the dominant process controlling the major ion composition of groundwater in the study area (Gibbs 1970).

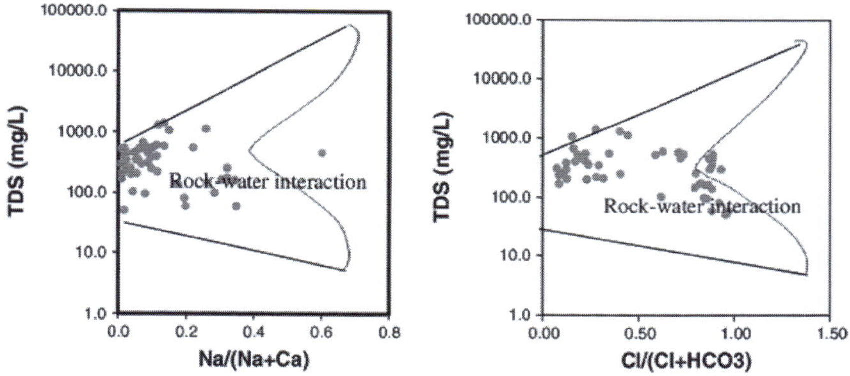

Fig.ure 5. Relation Na/(Na + Ca) vs TDS and between Cl/(Cl + HCO_3) vs TDS diagrams indicating water–rock interaction

The molar ratio of Na/Cl for groundwater samples of the study area generally ranges from 0.11 to 9.09 (Fig. 6a). Samples having Na/Cl ratio greater than one indicate excess sodium, which might have come from silicate weathering which is dominant in the aquifer materials of the study area. If silicate weathering is a probable source of sodium, the groundwater samples would have HCO_3^- as the most dominant abundant anion (Rogers 1989). This is because the reaction of feldspar minerals with carbonic acid in the presence of water releases HCO_3^- (Elango et al. 2003). Table 1 shows that HCO_3^- is the dominant anion in the groundwater of the study area; hence silicate weathering was the primary process responsible for the release of Na^+ into the groundwater. However, samples with a Na/Cl ratio around and less than one indicate the possibility of some other chemical processes, such as ion exchange (Elango et al.2003).

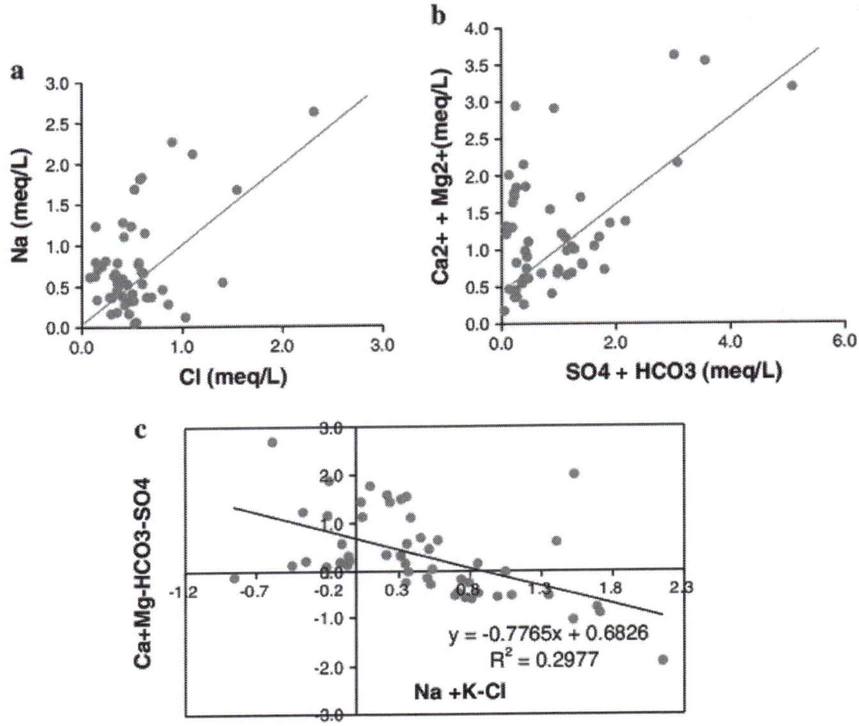

Figure 6. Graphs of different parameters: **a** Na vs. Cl, **b** Ca + Mg vs. $SO_4 + HCO_3$, **c** Ca + $Mg–HCO_3–SO_4$

Cerling et al. (1989) and Fisher and Mulican (1997) observed that the plot of $Ca^{2+} + Mg^{2+}$ vs $SO_4 + HCO_3{}^-$ is a major indicator to identify ion-exchange process. If ion exchange is the process, the points tend to shift to the right side of the plot due to an excess of $SO_4 + HCO_3{}^-$. If reverse ion exchange is the process, points tend to shift to the left side due to excess of Ca + Mg over $SO_4 + HCO_3{}^-$. The plot of the $Ca^{2+} + Mg^{2+}$ vs $SO_4 + HCO_3{}^-$ (Fig. 6b) in the study area shows that most of the groundwater samples of the alluvial aquifer are clustered around and above the 1:1 line, indicating reverse ion exchange (El-Sayed et al. 2012).

The evidence for cation exchange is also given by bivariate plot of $Ca^{2+} + Mg^{2+}–HCO_3–SO_4$ as a function of $Na^+ + K^+–Cl^-$, as shown in Fig. 6c. If cation exchange was not a significant reaction in the aquifer system, all data would plot close to the origin (McLean et al. 2000). But if cation exchange was an important composition controlling process, the relation between these two parameters should be linear with a slope of −1. Figure 6c indicates an increase in $Na^+ + K^+$ related to a decrease in $Ca^{2+} + Mg^{2+}$ or an increase in $HCO_3 + SO_4$. All data plot close to a straight line with a slope of −0.776, indicating that some of the Na^+, $Ca^{2+}$and Mg^{2+} participate in the ion-exchange reaction (Garcia et al. 2001). Furthermore, ion exchange and reverse ion exchange occurring in the groundwater were confirmed by two basic exchange indices, namely the chloro alkaline indices (CAI-1 and CAI-2):

$$\text{CAI-1} = [Cl^- - (Na^+ + K^+)]/Cl^- \tag{1}$$

$$\text{CAI-2} = [Cl^- - (Na^+ + K^+)]/(SO_4 + HCO_3 + CO_3 + NO_3) \tag{2}$$

(all values are expressed in meq/L). If there was ion exchange of Na^+ and K^+ from water with magnesium and calcium in the rock, then the exchange is known as direct when the indices are positive. If the exchange is in the reverse order, then the exchange is indirect and the indices are found to be negative (Schoeller 1967). The calculated chloro alkaline indices are presented in Table 1. It was observed that 75 % of the groundwater samples show negative ratios and thus exhibit a reverse ion exchange.

The area of study is characterised by shallow water table, underlain by very pyritic carbonaceous (peat, lignite and organic matter) wetland sediments. These sediments contain inorganic sulphides including pyrite and other iron-bearing minerals such as pyroxene, amphiboles, biotite, magnetite and olivine (Amadi et al. 1987). Oxygen transported into such layers, either by convection transport or direct exposure to air by lowering of groundwater levels results in oxidation of pyrite. Oxidation of pyrite, dissolution of these iron-bearing minerals and the decomposition of organic matter result in the release of acid, iron etc. into the soil and groundwater system (Jeong2001; Chirenje et al. 2007). In this study, the SO_4/Cl ratio is used as an indicator to illustrate potential pyritic oxidation. SO_4/Cl ratio ≥0.5 reflects pyrite oxidation. Table 1shows that in areas with SO_4/Cl ratios ≥0.5 also have high Fe concentrations in the groundwater due to oxidation of pyrite.

This results in a decrease in pH and thus explains in part the mild acidity and elevated concentration of iron observed in the groundwater system. Also, the low values of sulphate concentration in the groundwater (Table 1) are attributed to the removal of sulphate from the water by sulphate-reducing bacteria. The reduction of sulphate ions produces hydrogen sulphide; the unpleasant odour of which could easily be detected in the swamps of the Niger Delta (Amadi et al. 1987).

Dissolution and deposition

Mineral equilibrium calculations for groundwater are useful in predicting the presence of reactive minerals in the groundwater and estimating mineral reactivity (Deutsch1997). Using the saturation index (SI) approach, it is possible to predict the reactive mineralogy of the subsurface from groundwater without collecting the samples of the solid phase and analysing the mineralogy (Deutsch 1997). In the present study, the calculated values of SI for calcite and dolomite of the groundwater samples range from −2.99 to 2.0 and −3.89 to 3.64 with averages of −0.32 and −0.27 (Fig. 7a, b), respectively. The groundwater samples were kinetically undersaturated to oversaturated with respect to calcite and dolomite, and were undersaturated with respect to gypsum (Fig. 7c).

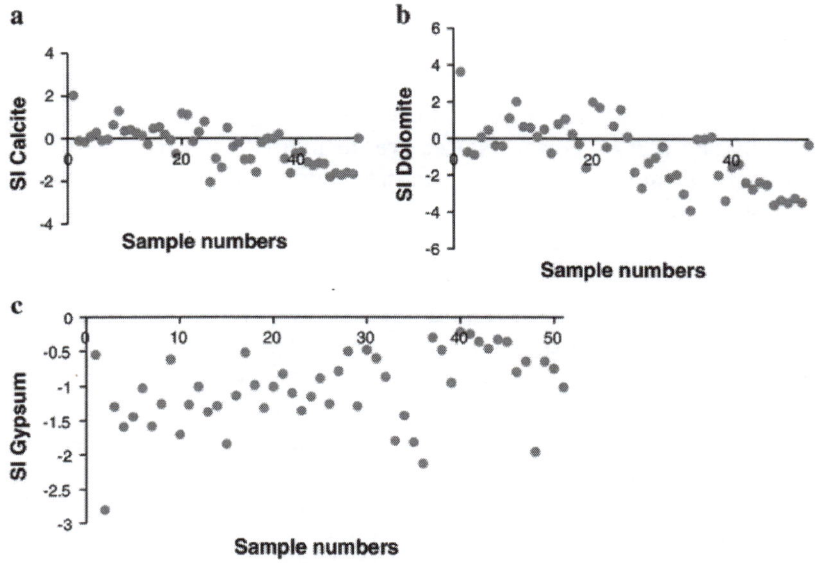

Figure 7. Saturation indices **a** Calcite, **b** Dolomite, **c** Gypsum

Principal component analysis

The results of the principal component analyses are shown in Table 3. They include the loading for the rotated component matrix, the eigenvalues for each component, and the percent and cumulative percent of variance explained by each component and communality, indicating the proportion of variance of each variable controlled by the set of components. From Table 3, it can be seen that the first three principal components together account for 78.38 % of the total variance in the dataset, in which the first principal component explains 44.09 % of the total variance, the second principal component exhibits 23.34 % of the total variance, and the third principal component explains 10.95 % of the total variance. The concentrations of EC, TDS, TA, TH, Na^+, K^+, Ca and HCO_3^- show high loadings (0.743–0.959), and the concentration of Cl^- has moderate positive loading (0.616) on the first principal component. On the second principal component, Fe and NO_3^- concentrations have high positive loadings (0.867–0.903), and the concentration of Mg^{2+} shows moderate positive loading (0.576). The combinations of Na^+, K^+, Ca, HCO_3^- and Cl^- ions, influencing the high positive loadings observed on the first principal component (Table 3), suggest that the first principal component is associated with a combination of various hydrogeochemical processes that contribute to enrich more mineralized water (TDS).

Table 3 .Rotated component matrix of chemical data of groundwater samples

Chemical variables	Principal components			Communality
	I	II	III	
pH	0.351	0.297	0.401	0.372
EC	0.959	0.068	0.084	0.931
TDS	0.954	0.096	0.096	0.929
TA	0.791	−0.260	0.249	0.755
TH	0.842	0.060	0.121	0.727
Na^+	0.843	−0.416	−0.074	0.890
K^+	0.829	−0.280	0.002	0.765
Ca^{2+}	0.842	0.250	−0.302	0.863
Mg^{2+}	0.489	0.576	−0.325	0.676
Fe	0.070	0.903	0.003	0.820
Cl^-	0.616	0.112	−0.609	0.764
SO_4^{2-}	−0.019	−0.178	−0.851	0.756
NO^{3-}	0.177	0.867	0.117	0.796
HCO_3^-	0.743	−0.531	0.254	0.899
Eigenvalues	6.613	3.502	1.642	
% of variance explained by component	44.085	23.344	10.946	
Cumulative % of variance	44.085	67.429	78.375	

For instance, a high positive loading of Na^+ and K^+ ions indicates natural weathering of rock minerals and various ion-exchange processes in the groundwater system. The strong loading of HCO_3^- ions with alkali and alkaline earth metals supports the view of natural weathering sources (Srivastava and Ramanathan 2008). The high positive loading of HCO_3^- is attributed to the reaction of feldspar minerals with carbonic acid in the presence of water (Elango et al. 2003). The concentrations of iron observed as high positive loadings on the second principal component (Table 3) are inferred to be related to the oxidation of inorganic sulphides including pyrite and leaching of iron-bearing minerals such as pyroxene, amphiboles, biotite, magnetite and olivine (Amadi et al. 1987). Nitrate has no significant lithologic source in the study area and it must be associated with the anthropogenic activities. Factor 3 explained that 10.95 % of the total variance is found to be associated with strong negative loading of SO_4^{2-} and weak pH. This factor may be attributed to anthropogenic activities such as domestic waste discharge.

CONCLUSION

The present study demonstrates the usefulness of factor analysis in combination with conventional graphical techniques in interpreting hydrogeochemical data to evaluate the major factors influencing groundwater composition. The degree to which the groundwater composition was influenced by various factors had been assessed by examining the factor scores. The hydrogeochemical characteristics of the groundwater revealed that the groundwater is slightly acidic and major anions and cations were below WHO prescribed limits for drinking water except iron. Prevalent water types in the study are: Ca–Cl, Ca–HCO_3–Cl, Ca–Na–Cl, Ca–Na–HCO_3–Cl, Ca–Na–HCO_3 and Na–HCO_3. The results of the multivariate statistical analysis in combination with conventional graphical methods indicate that the chemical composition of groundwater of the study area is strongly influenced by rock–water interaction, oxidation of inorganic sulphides including pyrite, weathering of silicates group of minerals enhanced by reverse ion-exchange processes. Weathering of silicate minerals controls the major ion chemistry of sodium, potassium, calcium and magnesium. The enhanced concentration of iron in the groundwater in the area is due to oxidation of pyrite and leaching of iron-bearing minerals. Nitrate and chloride have no significant lithologic source in the study area and must be associated with the anthropogenic activities such as sewage wastes.

ACKNOWLEDGMENTS

The authors are grateful to all home owners who allowed us access to their premises to collect water samples from their domestic boreholes. Special thanks to Mr. Suoye Diete-spiff of the Central Research Laboratory, Niger Delta University for the physico-chemical analysis and two anonymous reviewers whose comments improved the quality of this paper.

REFERENCES

1. Akpokodje EG (1986) A method of reducing the cement content of two stabilized Niger Delta soils. Q J Eng Geol London 19:359–363CrossRef
2. Amadi PA, Ofoegbu CO, Morrison T (1987) Hydrogeochemical assessment of groundwater quality in parts of the Niger Delta, Nigeria. Environ Geol Water Sci 14:195–202
3. Amajor LC (1991) Aquifers in the benin formation (Miocene-Recent), Eastern Niger Delta, Nigeria. lithostratigraphy, hydraulics and water quality. Environ Geol Water Sci 17(2):85–101
4. APHA (1995) Standard methods for the examination of water and waste water (17th ed) American Public Health Association, Washington
5. Ashley RP, Lloyd JW (1978) An example of the use of factor analysis and cluster analysis in groundwater chemistry interpretation. J Hydrogeol 39:355–364
6. Cerling TE, Pederson BL, Damm KLV (1989) Sodium–Calcium ion exchange in the weathering of shales: implications for global weathering budgets. Geology 17:552–554
7. Chirenje T, Epstein C, Mueller R (2007) Water quality issues in the outer coastal plains New Jersy. In: Sarkar et al. (eds) Developments in Environmental Science, 5, pp 561–589
8. Davis JC (1986) Statistics and data analysis in geology, 2nd edn. Willey, New York
9. Davis SN, De Wiest RJM (1966) Hydrogeology. Wiley, New York
10. Deutsch WJ (1997) Groundwater geochemistry: fundamentals and application to contamination. CRC, Boca Raton
11. Durotoye B (1989) Quaternary sediments in Nigeria: In Kogbe (ed.) Geology of Nigeria, Rock View (Nigeria) Ltd. Publishers, Jos
12. Elango L, Rannan R, Senthil KM (2003) Major ion chemistry and identification of hydrogeochemical processes of groundwater in a part of Kancheepuram district, Tamil Nadu, India. J Environ Geosci 10:157–166
13. El-Sayed MH, Moustafa M, El-Fadl MA, Shawky H (2012) Impact of hydrochemical processes on groundwater quality, Wadi Feiran, South Sinai, Egypt. Australian J Basic Appl Sci 6:638–654
14. Fisher RS, Mulican WF (1997) Hydrochemical evolution of sodium-sulfate and sodium-chloride groundwater beneath the Northern

Chihuahuan desert, Trans Pecos, Rexas, USA. Hydrogeol J 10:455–474

15. Garcia MG, del Hildalgo VM, Blessa MA (2001) Geochemistry of groundwater in the alluvial plain of Tucuman province, Argentina. Hydrogeol J 9:597–610

16. Gibbs RJ (1970) Mechanisms controlling World's water chemistry. Science 170:1088–1090

17. Guler C, Thyne GD, McCray JE, Turner AK (2002) Evaluation of graphical and multivariate statistical method for classification of water chemistry data. Hydrogeol J 10(4):455–474

18. Hem JD (1989) Study and interpretation of the chemical characteristics of natural water, 3rd edition. US Geol Surv Water Supply Paper 2254:263p

19. Jeong CH (2001) Mineral–water interaction and hydrogeochemistry in Samkwang mine area, Korea. Geochem J 35:1–12

20. Kogbe CA (1989) The cretaceous and paleogene sediments of southern Nigeria. In: Kogbe (ed.) Geology of Nigeria, Elizabethan Press, Lagos, pp 311–334

21. Krishnaraj S, Murugesan V, Vijayaraghaavan K, Sabarathinam C, Paluchamy A, Ramachandran M (2011) Use of hydrochemistry and stable isotopes as tools for groundwater evolution and contamination investigations. Geosciences 1:16–25

22. Lawrence FW, Upchurch SB (1982) Identification of recharge areas using geochemical factor analysis. Groundwater 20:680–687

23. Lee JY, Cheon JY, Lee KK, Lee SY, Lee MH (2001) Statistical evaluation of geochemical parameter distribution in a groundwater system contaminated with petroleum hydrocarbons. J Environ Qual 30:1548–1563

24. Lloyd JW, Heathcote JA (1985) Natural inorganic hydrochemistry in relation to groundwater, an introduction. Clarendon Press, Oxford

25. McLean W, Jankowski J, Levitt N (2000) Groundwater quality and sustainability in an alluvial aquifer, Australia. In: Sililo O (ed) Groundwater, past achievements and future chanllenges. Rotterdam, Balkenna, pp 567–573

26. Ofoegbu CO (1989) A review of the geology of the Benue Trough, Nigeria. J Afr Earth Sci 2:283–291

27. Okiongbo KS, Douglas RK (2013) Hydrogeochemical analysis and evaluation of groundwater quality in Yenagoa City and environs, Southern Nigeria. Ife J Sci 15:209–222

28. Okiongbo KS, Ogobiri G (2011) Geoelectric investigation of groundwater resources in parts of Bayelsa State, Nigeria. Res J Environ Earth Sci 3(6):620–624

29. Onyeagocha AC (1980) The petrology and depositional environment of the Benin Formation. Nigerian J Mining Geol 17:147–151

30. Reeder SW, Hitchon B, Levinson AA (1972) Hydrogeochemistry of the surface waters of the Mackenzie River drainage basin, Canada I. Factors controlling inorganic composition. Geochim Cosmochim Acta 36:825–865

31. Rogers RJ (1989) Geochemical comparison of groundwater in areas of New England, New York, and Pennsylvania. Groundwater 27:690–712

32. Ruiz F, Gomis V, Blasco P (1990) Application of factor analysis to the study of a coastal aquifer. Hydrogeol J 119:169–177

33. Schoeller H (1967) Qualitative evaluation of groundwater resources. In: Methods and techniques of groundwater investigation and development. Water Research, Series-*33:* UNESCO, pp 45–52

34. Short KC, Stauble AJ (1967) Outline of the geology of the Niger Delta. Bull AAPG 51:761–779

35. Srinivasamoorthy K, Chidambaram S, Prasanna MV, Vasanthavihar M, Peter J, Anandhan P (2008) Identification of major sources controlling groundwater chemistry from hard rock terrain—a case study from Mettur taluk, Salem district, Tamil Nadu, India. J Earth Syst Sci 117:49–58

36. Srivastava SK, Ramanathan AL (2008) Geochemical assessment of groundwater quality in vicinity of Bhalswa landfill, Delhi, India, using graphical and multivariate statistical methods. Environ Geol 53:1509–1528

37. World Health Organisation (WHO) (2004) Guidelines for drinking water quality: incorporating 1st and 2nd vol. 1 recommendations 3rd ed., Geneva

38. Wright JB, Hasting DA, Jones WB, Williams HR (1985) Geology and mineral resources of West Africa, Allen and Unwin Limited, UK, pp 107

9
CHAPTER

Exploration of Data Science Techniques to Predict Fatigue Strength of Steel From Composition and Processing Parameters

Ankit Agrawal[1*], Parijat D Deshpande[2], Ahmet Cecen[3], Gautham P Basavarsu[2], Alok N Choudhary[1] and Surya R Kalidindi[3,4]

[1]*Department of Electrical Engineering and Computer Science, Northwestern University, vanston, IL, USA.*

[2]*Tata Research Development and Design Centre, Tata Consultancy Services, Pune, Maharashtra, India.*

[3]*School of Computational Science and Engineering, Georgia Institute of Technology, Atlanta, GA, USA.*

4Woodruff School of Mechanical Engineering, Georgia Institute of Technology, Atlanta, GA, USA.

ABSTRACT

This paper describes the use of data analytics tools for predicting the fatigue strength of steels. Several physics-based as well as data-driven approaches have been used to arrive at correlations between various properties of alloys and their compositions and manufacturing process parameters. Data-driven approaches are of significant interest to materials engineers especially in arriving at extreme value properties such as cyclic fatigue, where the current state-of-the-art physics based models have severe limitations. Unfortunately, there is limited amount of documented success in these efforts. In this paper, we explore the application of different data science techniques, including feature selection and predictive modeling, to the fatigue properties of steels, utilizing the data from the

National Institute for Material Science (NIMS) public domain database, and present a systematic end-to-end framework for exploring materials informatics. Results demonstrate that several advanced data analytics techniques such as neural networks, decision trees, and multivariate polynomial regression can achieve significant improvement in the prediction accuracy over previous efforts, with R^2 values over 0.97. The results have successfully demonstrated the utility of such data mining tools for ranking the composition and process parameters in the order of their potential for predicting fatigue strength of steels, and actually develop predictive models for the same.

Keywords: Materials informatics Data mining Regression analysis Processing-property linkages

BACKGROUND

Causal relations are foundational to all advances in sciences and technology. In advancing materials science and engineering, the practitioners of the field have traditionally relied largely on observations made from cleverly designed controlled experiments and sophisticated physics-based models to establish the desired causal relations, e.g., process-structure-property (PSP) linkages. In recent vision-setting documents [1, 2], experts in materials science and engineering have identified data science and analytics as offering a third set of distinct tools (i.e., experiments, models, and data analytics making up the three foundational components of an integrated approach) for establishing the desired causal relations. Data science and analytics is expected to positively impact the ongoing materials development efforts by maximizing the accuracy and reliability of the core knowledge mined from large ensembles of (often heterogeneous and/or incomplete) datasets, and providing clear guidance for investment of future effort in as yet unexplored "white" spaces with the highest potential for success/benefit. In fact, data analytics techniques have already been successfully making inroads in the quest for new material design and discovery. The considerable interest and progress in the recent years has resulted in developing this new field and terming it as "Materials Informatics" [3, 4]. The Materials Genome Initiative [2] places a large emphasis on data-driven approaches. Progress in this direction is supplemented by the availability of large amounts of experimental and simulation data, enhanced computing tools and advances in data analytics, which is expected to augment rather than compete with existing analytics methods.

State-of-the-art data analytics

Over the last few decades, our ability to generate data has far exceeded our ability to make sense of it in practically all scientific domains, and materials science is no exception. This has led to the emergence of the fourth paradigm of science [5], which is data-driven science and discovery, and is based on developing predictive and discovery-based data mining approaches on big data in a comprehensive manner. Fourth paradigm compliments the three traditional scientific advancement models of mathematical modeling, experiments, and computer simulations. Indeed, the most advanced techniques in this field come from computer science, high-performance computing, machine learning and data mining algorithms, and via applications in business domain, climate science, bioinformatics, astronomy/cosmology, intrusion detection, network analysis and many others, where predictive data mining has been effectively used for decision making with significant gains in the outcomes relevant to that domain. For example, companies like Amazon [6, 7], Netflix [8], Google [9], Walmart [10] and Target [11] use predictive modeling for recommendations, personalized news, cost reductions, predicting demand and supply chain management at a massive scale providing lifts in sales and satisfaction. Scientists use predictive mining on big data to discover new stars/galaxies, predict hurricane paths, or predict structure of new materials. The accurate prediction of the path of hurricane Sandy illustrates an example of how the use and analysis of much larger data sets and algorithms can significantly improve accuracy.

The impressive advances made in the last two decades in both materials characterization equipment and in the physics-based multiscale materials modeling tools have ushered the BIG DATA age of materials science and engineering. With the advent of big data came the recognition that advanced statistics and modern data analytics would have to play an important role in the future workflows for the development of new or improved materials.

Several case studies illustrating the potential benefits of this emerging new field of Materials Informatics (MI) have already been published in literature. Rajan et al. [12] applied principal component analysis (PCA) on a database consisting of 600 compounds of high temperature superconductors to identify patterns and factors which govern this important property. They observed that the dataset clusters according to the average valency, a criterion which has been reported in literature to be of utmost importance for superconducting property. They concluded that informatics techniques allow one to investigate complex multivariate information in an accelerated and yet physically meaningful manner. Suh and Rajan [13] applied informatics tools on a dataset consisting of AB2N4 spinel nitrides to find the statistical interdependency of factors that may influence chemistry-structure-property relationships. Using partial least squares (PLS), they developed a quantitative structure-activity relationship (QSAR) relating bulk modulus of AB2N4 spinels with a variety of control parameters. They observed a strong agreement between the properties predicted based on ab-initio calculations and the ones based strictly on a data-driven approach. Nowers et al. [14] investigated property-structure-processing relations during interpenetrating polymer network (IPN) formation in epoxy/acrylate systems using an informatics approach. They concluded that material informatics is a very efficient tool which can be utilized for additional materials development. Gadzuric et al. [15] applied informatics tools on molten salt database to predict enthalpy (δH_{form}) and Gibbs free energy of formation (δG_{form}) of lanthanide halides. The results of the analysis indicated a high level of confidence for the predictions. George et al. [16] applied similar approach on a dataset consisting of binary and ternary metal hydrides to investigate the interrelationships among material properties of hydrides. They developed a relationship between entropy of a hydride and its molar volume which was in close agreement with the theoretical predictions. Singh et al. [17] developed a neural network model in which the yield and tensile strength of the steel was estimated as a function of some 108 variables, including the chemical composition and an array of rolling parameters. Fujii et al. [18] applied neural network approach for prediction of fatigue crack growth rate of nickel base superalloys. They modeled the rate as a function of 51 variables and demonstrated the ability of such methods for investigation of new phenomena in cases where the information cannot be accessed experimentally. Hancheng et al. [19] developed an adaptive fuzzy neural network model to predict strength based on compositions and microstructure. Runway stiffness prediction and evaluation models have also been developed using techinques such as genetic programming [20] and artificial neural networks [21]. Wen et al. [22] applied support vector regression (SVR) approach for prediction of corrosion rate of steels under different seawater environments. They concluded that SVR is a promising and practical method for real-time corrosion tracking of steels. Rao et al. [23] applied SVR for prediction of grindability index of coal and concluded that SVR is a promising technique and needs smaller data set for training the model than artificial neural network (ANN) techniques. To the best of our knowledge, there is only one prior study [24] dealing with fatigue strength prediction using the NIMS database (same data that we use in this work; details of the data provided later). It applied PCA on the data and subsequently performed partial least square regression (PLSR) on the different clusters identified by PCA for making predictions. Large R^2 values ranging between 0.88 and 0.94 were obtained for the resulting clusters.

Motivation

The prior MI case studies cited above have established the unequivocal potential of this emerging discipline in accelerating discovery and design of new/improved materials. However, there still does not exist a standardized set of protocols for exploring this approach in a systematic manner on many potential applications, and thus, establishing the composition-processing-structure-property relationships still remains an arduous task. A report published by NRC [1] stated that materials design has not been able to keep pace with the product design and development cycle and that insertion of new materials has become more infrequent. This poses a threat to industries such as automotive and aerospace, in which the synergy between product design, materials, and manufacturing is a competitive advantage.

In this paper, we embark on establishing a systematic framework for exploring MI, and illustrate it by establishing highly reliable causal linkages between process variables in a class of steels, their chemical compositions, and their fatigue strengths. The approach described in this work comprises of four main steps: (i) Preprocessing for consistent description of data, which can include things like filling in missing data wherever possible, with the help of appropriate domain knowledge; ii) Feature selection for attribute ranking and/or identifying the best subset of attributes for establishing a given linkage; (iii) Predictive modeling using multiple statistical and advanced data-driven strategies for the establishment of the desired linkages, (iv) Critical evaluation of the different informatics approaches using appropriate metrics and evaluation setting to avoid model over-fitting.

Accurate prediction of fatigue strength of steels is of particular significance in materials science to several advanced technology applications because of the extremely high cost (and time) of fatigue testing and often debilitating consequences of fatigue failures. Fatigue strength is the most important and basic data required for design and failure analysis of mechanical components. It is reported that fatigue accounts for over 90% of all mechanical failures of structural components [25]. Hence, fatigue life prediction is of utmost importance to both the materials science and mechanical engineering communities. The unavailability of recorded research in using a large number of heat treatment process parameters, composition to predict extreme value properties such as fatigue strength has led us to work on this problem. The complex interaction between the various input variables have baffled the conventional attempts in pursuing this work and advanced data analytics techniques may lead the path. The aim of this study is thus to fill some of the gaps encountered in this work and serve as a preliminary guide for prospective researchers in this field. The scope of this paper includes application of a range of machine learning and data analytics methods applied for the problem of fatigue strength prediction of steels using composition and processing parameters. A conference version of this paper with preliminary results appeared in Proceedings of the 2nd World Congress on Integrated Computational Materials Engineering (ICME 2013) [26].

Data

Fatigue Dataset for Steel from National Institute of Material Science (NIMS) MatNavi [27] was used in this work, which is one of the largest databases in the world with details on composition, mill product (upstream) features and subsequent processing (heat treatment) parameters. The database comprises carbon and low-alloy steels, carburizing steels and spring steels. Fatigue life data, which pertain to rotating bending fatigue tests at room temperature conditions, was the target property for which we aimed to construct predictive models in the current study. The features in the dataset can be categorized into the following:

- Chemical composition - %C, %Si, %Mn, %P, %S, %Ni, %Cr, %Cu, %Mo (all in wt. %)
- Upstream processing details - ingot size, reduction ratio, non-metallic inclusions
- Heat treatment conditions - temperature, time and other process conditions for normalizing, through-hardening, carburizing-quenching and tempering processes
- Mechanical properties - YS, UTS, %EL, %RA, hardness, Charpy impact value (J/cm2), fatigue strength.

The data used in this work has 437 instances/rows, 25 features/columns (composition and processing parameters), and 1 target property (fatigue strength). The 437 data instances include 371 carbon and low alloy steels, 48 carburizing steels, and 18 spring steels. This data pertains to various heats of each grade of steel and different processing conditions. The details of the 25 features and given in Table 1.

METHODS

The overall proposed approach is illustrated in Figure 1. The raw data is preprocessed for consistency using domain knowledge. Ranking-based feature selection methods are also used to get an idea of the relative predictive potential of the attributes. Different regression-based predictive modeling methods are then used on the preprocessed and/or transformed data to construct models to predict the fatigue strength, given the composition and processing parameters. All constructed models are evaluated using Leave-One-Out Cross Validation with respect to various metrics for prediction accuracy. Below we present the details of each of the 4 stages.

Table 1. NIMS data features

Abbreviation	Details
C	% Carbon
Si	% Silicon
Mn	% Manganese
P	% Phosphorus
S	% Sulphur
Ni	% Nickel
Cr	% Chromium
Cu	% Copper
Mo	% Molybdenum
NT	Normalizing Temperature
THT	Through Hardening Temperature
THt	Through Hardening Time
THQCr	Cooling Rate for Through Hardening
CT	Carburization Temperature
Ct	Carburization Time
DT	Diffusion Temperature
Dt	Diffusion time
QmT	Quenching Media Temperature (for Carburization)
TT	Tempering Temperature
Tt	Tempering Time
TCr	Cooling Rate for Tempering
RedRatio	Reduction Ratio (Ingot to Bar)
dA	Area Proportion of Inclusions Deformed by Plastic Work
dB	Area Proportion of Inclusions Occurring in Discontinuous Array
dC	Area Proportion of Isolated Inclusions
Fatigue	**Rotating Bending Fatigue Strength (10^7 Cycles)**

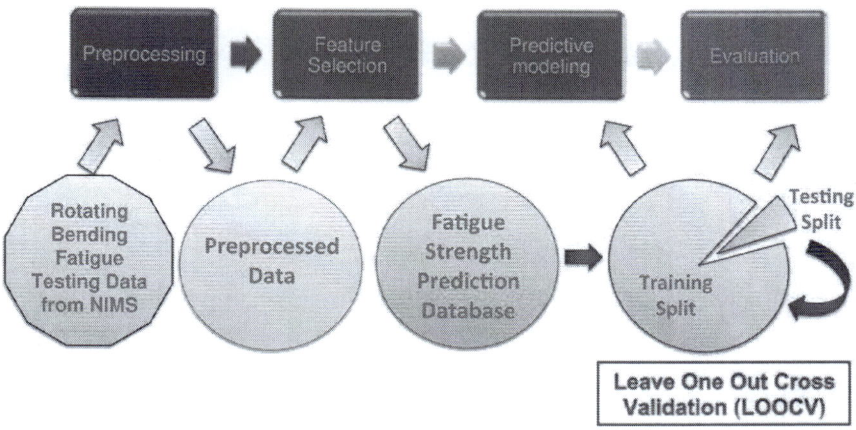

Figure 1.Block diagram. Schematic of overall proposed approach to fatigue strength prediction of steels with knowledge discovery and data mining.

Preprocessing

Understanding and cleaning the data for proper normalization is one of the most important steps for effective data mining. Appropriate preprocessing, therefore, becomes extremely crucial in any kind of predictive modeling, including that of fatigue strength. The dataset used in this study consists of multiple grades of steel and in some records, some of the heat treatment processing steps did not exist. In particular, different specimens are subjected to different heat treatment conditions. For example, some are normalized and tempered, some are through hardened and tempered, and others are carburized and tempered. There could be cases where normalization is done prior to carburization and tempering. In order to bring in a structure to the database, we have included all the key processes in the data-normalization, through hardening, carburization, quenching and tempering. For the cases where the actual process does not take place, we set the appropriate duration/time variable to zero with corresponding temperature as the austenization temperature or the average of rest of the data where the process exists. Setting the time to zero would essentially mean that no material transformation occurs. An artifact of our resulting data is that we are treating temperature and time as independent variables whereas they actually make sense only when seen together.

Feature selection

Information gain

This is an entropy-based metric that evaluates each attribute independently in terms of its worth by measuring the information gain with respect to the target variable:

$$IG(Class, Attrib) = H(Class) - H(Class|Attrib) \quad (1)$$

where $H(.)$ denotes the information entropy. The ranking generated by this method can be useful to get insights about the relative predictive potential of the input features.

SVD-PCA

Singular value decomposition is a matrix factorization defined as:

$$D = U \times S \times V \quad (2)$$

where, D is the data matrix such that every observation is represented by a row and each column is an explanatory variable, U is the matrix of left singular vectors, V is the matrix of right singular vectors and S is the diagonal matrix of singular values. In this case, $A=U\times S$ is a transformation of D where the data is represented by a new set of explanatory variables such that each variable is a known linear combination of the original explanatory parameters. The dimensions of A are also referred to as the Principal Components (PC) of the data.

Predictive modeling

We experimented with 12 predictive modeling techniques in this research study, which include the following:

Linear regression

Linear regression probably the oldest and most widely used predictive model, which commonly represents a regression that is linear in the unknown parameters used in the fit. The most common form of linear regression is least squares fitting [28]. Least squares fitting of lines and polynomials are both forms of linear regression.

Pace regression

It evaluates the effect of each feature and uses a clustering analysis to improve the statistical basis for estimating their contribution to overall regression. It can be shown that pace regression is optimal when the number of coefficients tends to infinity. We use a version of Pace Regression described in [29, 30].

Regression post non-linear transformation of select input variables

A non-linear transformation of certain input variables can be done and the resulting data-set used for linear regression. In this study, the temperature variation effects on the diffusion equation are modelled according to the Arrhenius' empirical equation as $exp(-1/T)$ where T is measured in Kelvin.

Robust fit regression

The robust regression method [31] attempts to mitigate the shortcomings which are likely to affect ordinary linear regression due to the presence of outliers in the data or non-normal measurement errors.

Multivariate polynomial regression

Ordinary least squares (OLS) regression is governed by the equation:

$$\beta = (X'X)^{-1}X'Y \qquad (3)$$

where β is the vector of regression coefficients, X is the design matrix and Y is the vector of responses at each data point. Multivariate Polynomial Regression (MPR) is a specialized instance of multivariate OLS regression that assumes that the relationship between regressors and the response variable can be explained with a standard polynomial. Standard polynomial here refers to a polynomial function that contains every polynomial term implied by a multinomial expansion of the regressors with a given degree (sometimes also referred to as a polynomial basis function). Polynomials of various degrees and number of variables are interrogated systematically to find the most suitable fit. There is a finite number of possible standard polynomials that can be interrogated due to the degree of freedom imposed by a particular dataset; the number of terms in the polynomial (consequently the number of coefficients) cannot exceed the number of data points.

Instance-based

This is a lazy predictive modeling technique which implements the K-nearest-neighbour (kNN) modeling. It uses normalized Euclidean distance to find the training instance closest to the given test instance, and predicts the same class as this training instance [32]. If multiple instances have the same (smallest) distance to the test instance, the first one found is used. It eliminates the need for building models and supports adding new instances to the training database dynamically. However, the zero training time comes at the expense of a large amount of time for testing since each test instance needs to be compared with all the data instances in the training data.

KStar

KStar [33] is another lazy instance-based modeling technique, i.e., the class of a test instance is based upon the class of those training instances similar to it, as determined by some similarity function. It differs from other instance-based learners in that it uses an entropy-based distance function. The underlying technique used of summing probabilities over all possible paths is believed to contribute to its good overall performance over certain rule-based and instance-based methods. It also allows an integration of both symbolic and real valued attributes.

Decision table

Decision table is a rule-based modeling technique that typically constructs rules involving different combinations of attributes, which are selected using an attribute selection search method. It thus represents one of the simplest and most rudimentary ways of representing the output from a machine learning algorithm, showing a decision based on the values of a number of attributes of an instance. The number and specific types of attributes can vary to suit the needs of the task. Simple decision table majority classifier [34] has been shown to sometimes outperform state-of-the-art classifiers. Decision tables are easy for humans to understand, especially if the number of rules are not very large.

Support vector machines

SVMs are based on the Structural Risk Minimization (SRM) principle from statistical learning theory. A detailed description of SVMs and SRM is available in [35]. In their basic form, SVMs attempt to perform classification by constructing hyperplanes in a multidimensional space that separate the cases of different class labels. It supports both classification and regression tasks and can handle multiple continuous and nominal variables. Different types of kernels can be used in SVM models, including linear, polynomial, radial basis function (RBF), and sigmoid. Of these, the RBF kernel is the most recommended and popularly used, since it has finite response across the entire range of the real x-axis.

Artificial neural networks

ANNs are networks of interconnected artificial neurons, and are commonly used for non-linear statistical data modeling to model complex relationships between inputs and outputs. The network includes a hidden layer of multiple artificial neurons connected to the inputs and outputs with different edge weights. The internal edge weights are 'learnt' during the training process using techniques like back propagation. Several good descriptions of neural networks are available [36, 37].

Reduced error pruning trees

A Reduced Error Pruning Tree (REPTree) [38] is an implementation of a fast decision tree learner. A decision tree consists of internal nodes denoting the different attributes and the branches denoting the possible values of the attributes, while the leaf nodes indicate the final predicted value of the target variable. REPTree builds a decision/regression tree using information gain/variance and prunes it using reduced-error pruning. In general, a decision tree construction begins at the top of the tree (root node) with all of the data. At each node, splits are made according to the information gain criterion, which splits the data into corresponding branches. Computation on remaining nodes continues in the same manner until one of the stopping criterions is met, which include maximum tree depth, minimum number of instances in a leaf node, minimum variance in a node.

M5 model trees

M5 Model Trees [39] are a reconstruction of Quinlan's M5 algorithm [40] for inducing trees of regression models, which combines a conventional decision tree with the option of linear regression functions at the nodes. It tries to partition the training data using a decision tree induction algorithm by trying to minimize the intra-subset variation in the class values down each branch, followed by back pruning and smoothing, which substantially increases prediction performance. It also uses the techniques used in CART [41] to effectively deal with enumerated attributes and missing values.

Evaluation

Traditional regression-based methods such as linear regression are typically evaluated by building the model (a linear equation in the case of linear regression) on the entire available data, and computing prediction errors on the same data. Although this approach works well in general for simple regression methods, it is nonetheless susceptible to over-fitting, and thus can give over-optimistic accuracy numbers. In particular, a data-driven model can, in principle learn every single instance of the dataset and thus result in 100% accuracy on the same data, but will most likely not be able to work well on unseen data. For this reason, advanced data-driven techniques that usually result in black-box models need to be evaluated on data that the model has not seen while training. A simple way to do this is to build the model only on random half of the data, and use the remaining half for evaluation. This is called the train-test split setting for model evaluation. Further, the training and testing halves can then also be swapped for another round of evaluation and the results combined to get predictions for all the instances in the dataset. This setting is called 2-fold cross validation, as the dataset is split into 2 parts. It can further be generalized to k-fold cross validation, where the dataset is randomly split into k parts. $k-1$ parts are used to build the model and the remaining 1 part is used for testing. This process is repeated k times with different test splits, and the results combined to get preductions for the all the instances in the dataset using a model that did not see them while training. Cross validation is a standard evaluation setting to eliminate any chances of over-fitting. Of course, k-fold cross validation necessitates builing k models, which may take a long time on large datasets.

Leave-one-out cross validation

We use leave-one-out cross validation (LOOCV) to evaluate and compare the prediction accuracy of the models. LOOCV is commonly used for this purpose particularly when the dataset is not very large. It is a special case of the more generic k-fold cross validation, with $k=N$, the number of instances in the dataset. The basic idea here is to estimate the accuracy of the predictive model on unseen input data it may encounter in the future, by withholding part of the data for training the model, and then testing the resulting model on the withheld data. In LOOCV, to predict the target attribute for each data instance, a separate predictive model is built using the remaining $N-1$ data instances. The resulting N predictions can then be compared with the N actual values to calculate various quantitative metrics for accuracy. In this way, each of the N instances is tested using a model that did not see it while training, thereby maximally utilizing the available data for model building, and at the same time eliminating the chances of over-fitting of the models.

Evaluation metrics

Quantitative assessments of the degree to how close the models could predict the actual outputs are used to provide an evaluation of the models' predictive performances. A multi-criteria assessment with various goodness-of-fit statistics was performed using all the data vectors to test the accuracy of the trained models. The criteria that are employed for evaluation of models' predictive performances are the coefficient of correlation (R), explained variance (R^2), Mean Absolute Error (MAE), and Root Mean Squared Error ($RMSE$), Standard Deviation of Error (SDE) between the actual and predicted values. The last three metrics were further normalized by the actual fatigue strength values to express them as error fractions. The definitions of these evaluation criteria are as follows:

$$R = \frac{\sum_{i=1}^{N}(y_i - \overline{y})(\hat{y}_i - \overline{\hat{y}})}{\sqrt{\sum_{i=1}^{N}(y_i - \overline{y})^2 \sum_{i=1}^{N}(\hat{y}_i - \overline{\hat{y}})^2}} \tag{4}$$

$$MAE = \overline{e} = \frac{1}{N}\sum_{N}|y - \hat{y}| \tag{5}$$

$$RMSE = \sqrt{\frac{1}{N}\sum_{N}(y - \hat{y})^2} \tag{6}$$

$$SDE = \sqrt{\frac{1}{N}\sum_{N}(|y - \hat{y}| - \overline{e})^2} \tag{7}$$

$$MAE_f = \overline{e_f} = \frac{1}{N}\sum_{N}\left|\frac{y - \hat{y}}{y}\right| \tag{8}$$

$$RMSE_f = \sqrt{\frac{1}{N}\sum_{N}\left(\frac{y - \hat{y}}{y}\right)^2} \tag{9}$$

$$SDE_f = \sqrt{\frac{1}{N}\sum_{N}\left(\left|\frac{y - \hat{y}}{y}\right| - \overline{e_f}\right)^2} \tag{10}$$

where y denotes the actual fatigue strength values (MPa), $\hat{y}\hat{y}$ denotes the predicted fatigue strength values (MPa), and N is the number of instances in the dataset.

The square of the coefficient of correlation, R^2, represents the variance explained by the model (higher the better), and is considered one of the most important metrics for evaluating the accuracy of regressive prediction models. Another useful metric is the fractional mean absolute error, MAE_f, which represents the error rate (lower the better).

RESULTS AND DISCUSSION

We used the following statistical and data mining software for conducting the analysis reported in this paper: R [42], MATLAB [43], WEKA [44]. Default parameters were used unless stated otherwise.

The entire available data-set was assessed for visible clustering by employing K-means clustering technique. The cluster plot demonstrates inherent clustering in the available data, which agrees with the a priori knowledge of the dataset. The distinct clustering in the available data represents 3 clusters according to the grade of steels as depicted in Figure 2. These clusters however do not offer sufficient data-points to create individual meta-models for each cluster and hence, for all methods used, the entire data-set is used to develop predictive models.

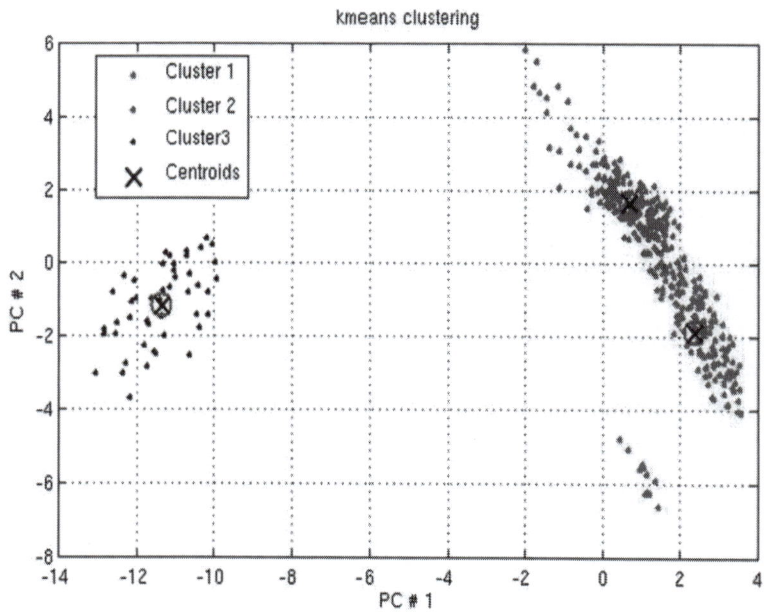

Figure 2. Cluster visualization. K-means clustering of the NIMS data using the top 2 principal components with K=3, clearly depiciting the 3 clusters corresponding to the 3 steel grades.

The information gain metric was calculated for each of the 25 input attributes. For this purpose, the numerical fatigue strength was discretized into 5 equal-width bins as this only works for categorical target attributes. The relative predictive power of the 25 input attributes is shown in Figure 3. All the attributes were retained for building various predictive models as all of them were found to have significant predictive potential. We also looked at the correlation values of the 25 input features with the fatigue strength, as shown in Figure 4. Interestingly, TT (tempering temperature) shows up as the most important attribute for predicting fatigue strength in Figure 3. This is because the dataset consists of multiple grades of steel, each with a narrow yet significantly different range of TT. For example, TT for Through-hardened-tempered (without carburization) is around 400°C and that with carburization is around 200°C. These two situations will lead to a large difference in the fatigue strength. Thus, there is no surprise that the influence of TT seems high. However, the truth is that having a carburization step is what makes the key difference in the fatigue strength. Nevertheless, tempering will have a significant effect and this is reflected by the influence of tempering time in Figure 4. Figure 4 also identifies other variables such as carburizing temperature or through hardening temperature as important influencing factors. These are in line with expected results.

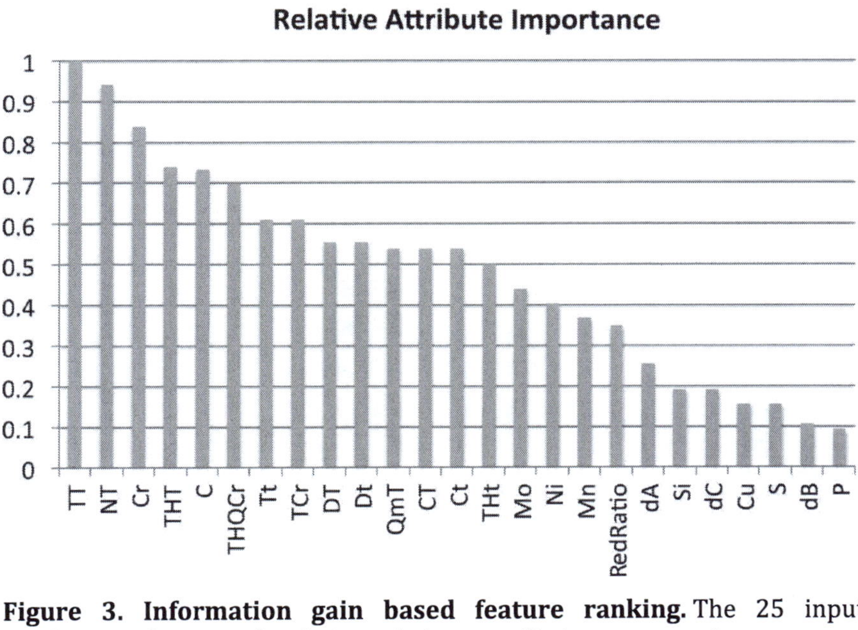

Figure 3. Information gain based feature ranking. The 25 input attributes along with their relative predictive power in terms of information gain.

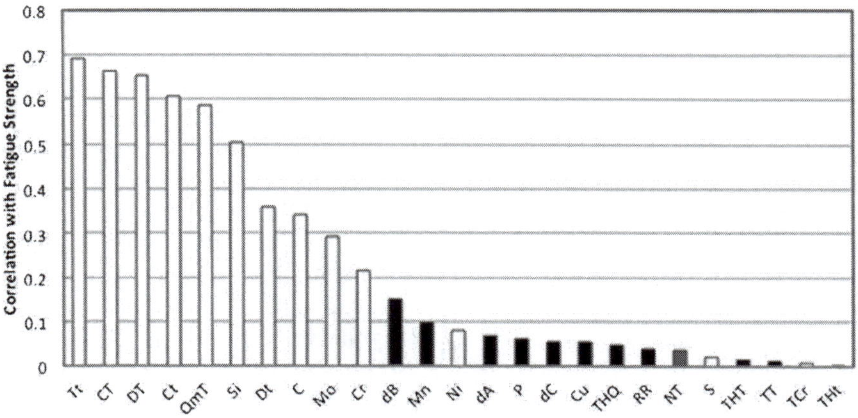

Figure 4. Correlation based feature ranking. The 25 input attributes along with their correlation values with fatigue strength. White (black) bars indicate positive (negative) influence.

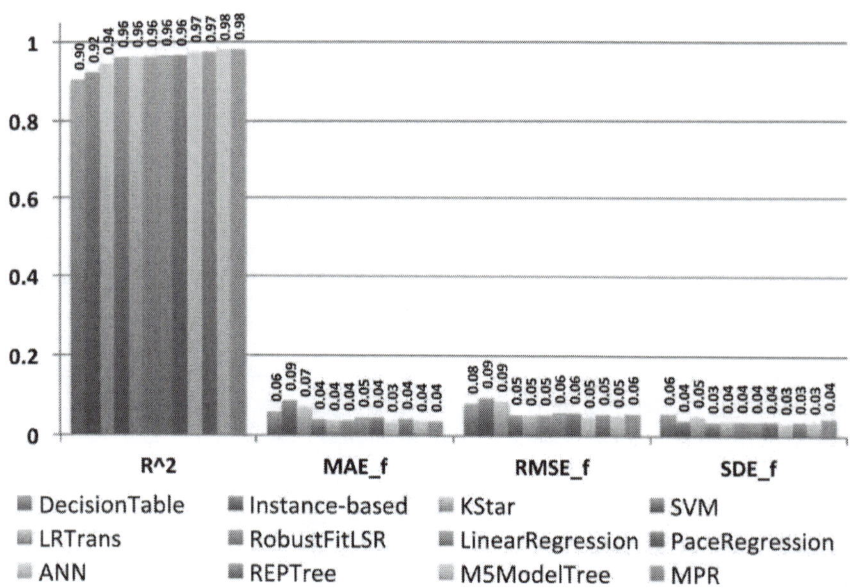

Figure 5. Results comparison. R^2, MAE_f, $RMSE_f$, and SDE_f for 12 predictive modeling techniques. Better fitting accuracy is characterized by higher values of R^2 and lower values of MAE_f, $RMSE_f$, and SDE_f.

As mentioned before, we use Leave-One-Out Cross Validation (LOOCV) for model evaluation. Figure 5 and Table 2present the LOOCV prediction accuracy of the 12 modeling techniques used in this work, in terms of the metrics discussed earlier. Clearly, many of the employed data analytics techniques are able to achieve a high predictive accuracy, with R^2 values ~0.98, and error rate <4%. This is extremely encouraging since it significantly outperforms the only prior study on fatigue strength prediction [24], which reported R^2 values of <0.94. It is well known in the field of predictive data analytics that it becomes progressively more and more challenging to increase the accuracy of prediction beyond a certain point. To put it in context of this study, an increase in R^2 from 0.94 to 0.98 should not be viewed as simply an improvement of 0.04 or 4%. Rather, it should be seen with respect to the available scope for improvement of 0.06 (= 1.00 - 0.94). Thus, a more reasonable evaluation of the improvement accomplished by the current study over prior work would be about 66% (0.04/0.06), which is very significant.

Table 2. Results comparison

Method	*R*	R^2	*MAE*	*RMSE*	*SDE*	MAE_f	$RMSE_f$	SDE_f
DecisionTable	0.9494	0.9014	34.8762	58.5932	47.1371	0.0584	0.0806	0.0557
IBk	0.9589	0.9195	46.0320	53.2749	26.8499	0.0859	0.0940	0.0382
KStar	0.9702	0.9413	36.9986	45.3779	26.3029	0.0706	0.0857	0.0487
SVM	0.9795	0.9594	24.2820	37.6250	28.7736	0.0400	0.0530	0.0349
LRTrans	0.9796	0.9596	22.3336	37.4748	30.1272	0.0370	0.0514	0.0357
RobustFitLSR	0.9804	0.9612	22.2152	37.2188	29.8960	0.0369	0.0520	0.0366
LinearRegression	0.9815	0.9633	25.6006	35.7168	24.9345	0.0456	0.0581	0.0360
PaceRegression	0.9816	0.9635	25.0302	35.5733	25.3065	0.0439	0.0565	0.0356
ANN	0.9861	0.9724	19.7778	31.0545	23.9695	0.0343	0.0470	0.0322
REPTree	0.9862	0.9726	22.5671	30.9401	21.1907	0.0414	0.0542	0.0349
M5ModelTree	0.9890	0.9781	19.3760	27.6065	19.6870	0.0353	0.0484	0.0332
MPR	0.9900	0.9801	18.5529	26.4378	18.8563	0.0350	0.0556	0.0432

Figure 6 presents the scatter plots for the 12 techniques. As can be seen from these plots, the three grades of steels are well separated in most of the techniques, and different techniques tend to perform better for different grades. Figure 7 shows the histograms of the error fractions for each of the techniques, to visualize the spread in the prediction errors. As expected, the spread in the error reduces as R^2 values improve. However, a point to be noted is that even though R^2 value is high, there are regions of data clusters where data fit is not sufficiently high and this is reflected in the nature of distribution of errors. Thus, the methods that result in bimodal distribution of errors or the ones with significant peaks in higher error regions are not so good even though their reported R^2 may be reasonable.

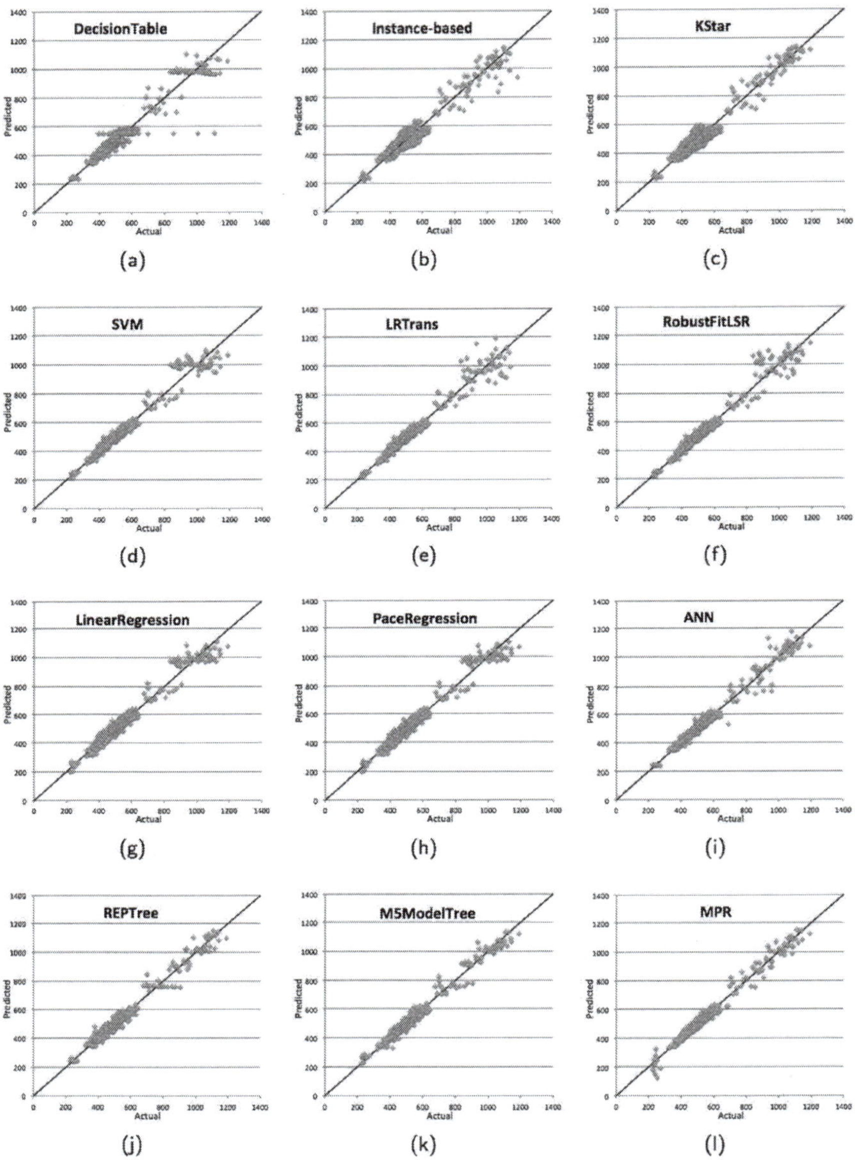

Figure 6. Scatter plots. Scatter plots for the 12 modeling techniques. X-axis and Y-axis denote the actual and predicted fatigue strength (in MPa) respectively. **a)** Decision Table; **b)** Instance-based; **c)** KStar; **d)** Support Vector Machines; **e)** Regression with Transformed Terms; **f)** RobustFit Regression; **g)** Linear Regression; **h)**Pace Regression; **i)** Artificial Neural Networks; **j)** Reduced Error Pruning Trees; **k)** M5 Model Trees; **l)**Multivariate Polynomial Regression.

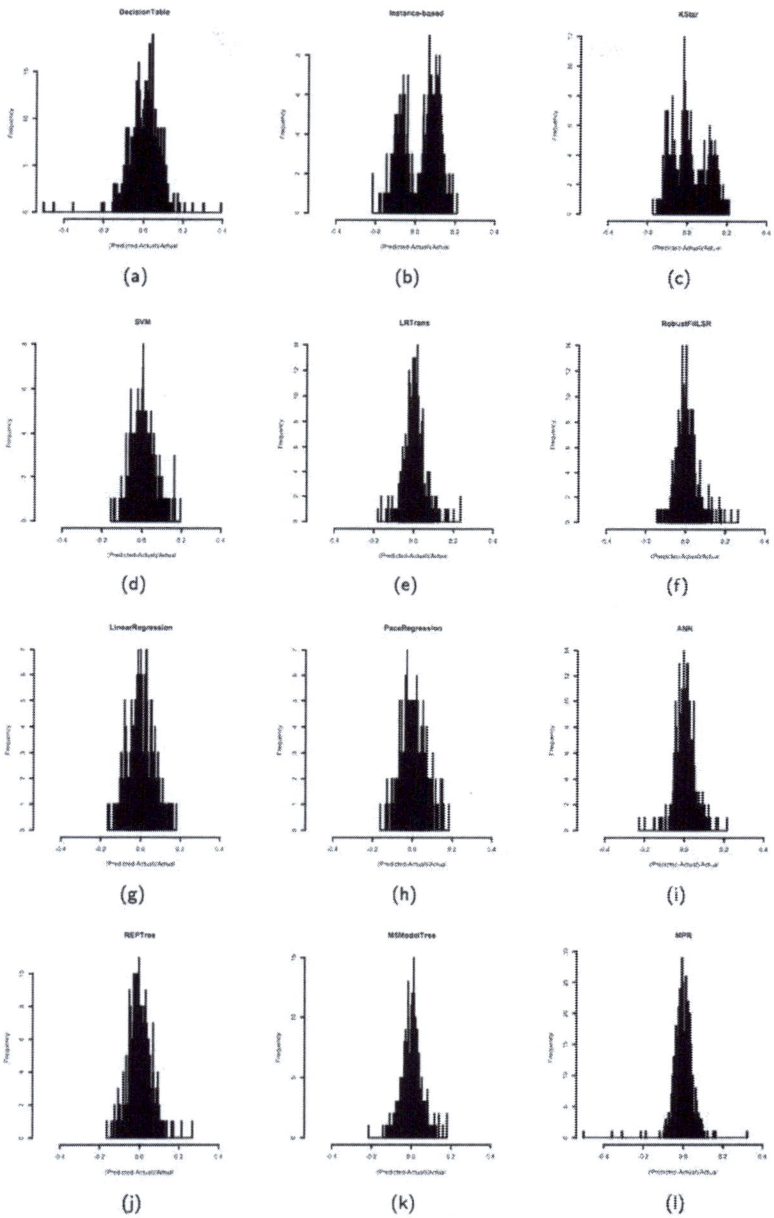

Figure 7. Error histograms. Histograms of the error fractions for the 12 modeling techniques. X-axis and Y-axis denote the error fraction bins and the frequency respectively. **a)** Decision Table; **b)** Instance-based; **c)**KStar; **d)** Support Vector Machines; **e)** Regression with Transformed Terms; **f)** RobustFit Regression; **g)**Linear Regression; **h)** Pace Regression; **i)** Artificial Neural Networks; **j)** Reduced Error Pruning Trees; **k)** M5 Model Trees; **l)** Multivariate Polynomial Regression.

The general opinion in data mining community about predictive modeling is that it is more helpful to know about a set of well performing techniques for a given problem rather than identifying a single winner. We have thus examined 12 different techniques for predictive modeling of fatigue strength, and it is shown that a number of different approaches produce highly reliable linkages. In particular, neural networks, decision trees, and multivariate polynomial regression were found to achieve a high R^2 value of greater than 0.97, which is significantly better than what has been previously reported in the literature. This is also shown by narrow distribution of errors. It is very encouraging to see that despite the limited amount of data available in this dataset, the data-driven analytics models were able to achieve a reasonably high degree of accuracy.

Although the main contribution of this paper is to present an end-to-end framework for exploring predictive materials informatics, and its application on NIMS data is a specific example of the application of the framework, it is nonetheless important for completeness to discuss some of the limitations of the proposed framework's specific application on the NIMS dataset. Since the data used in this study is very small compared to the typical amounts of data used in data mining studies in other domains, we believe that the obtained high accuracy is but an encouragement to use more data (possibly combine data from heterogenous sources) to further validate the results and/or making the model more robust. One possibility would be to add structure information to the data, which may ease the application of the developed models to actionable materials design, as structure information is what is primary responsible for the resulting properties. Another limitation of the NIMS data used in this study is the significantly different number of data instances corresponding to the different types of steels. Hence the predictive models, which are developed over the entire data may not be highly accurate for all steel types, which is also evident from the scatter plots. Possible approaches to deal with this imbalanced data distribution are discussed in the next section.

CONCLUSIONS

Materials Informatics, steeped in modern data analytics and advanced statistics, is fast emerging as a key enabler for accelerated and cost-effective development of new and improved materials targeted for advanced technologies. One of the core challenges addressed by this nascent field is the successful mining of highly reliable, quantitative, linkages capturing the salient connections between chemical compositions, processing history, and the final properties of the produced material. These linkages can provide valuable guidance to future effort investment with tremendous potential for cost-savings.

In this paper, we have tried to critically explore the viability of extracting such linkages from open access databases. As a specific example, we have focused on extracting reliable linkages between chemical compositions, processing history, and fatigue strength of a class of steels using data available from the open access materials database hosted by Japan's National Institute for Materials Science (NIMS). In this study, a range of advanced data analytics techniques, typically involving a combination of feature selection and regression methods, have been successfully employed and critically evaluated for the problem of fatigue strength prediction of different grades of steels.

There are several directions of future work that can stem from the present research. From the data analytics point of view, ensemble predictive modeling can be used to combine the results from multiple predictive models using same and/or different techniques built on different random subsets of the training data. Apart from this, since the scatter plots showed that different techniques can work well for different steel types, we can also try hierarchical predictive modeling, where we first try to classify the input test instance into one of the three grades of steel, and subsequently use the appropriate model(s) for that grade to predict the fatigue strength. From the materials science point of view, it would be good to explore the use of additional input features that may be easily measurable like some mechanical properties. Methods for using grouped variables representing each processing step could be of significant utility as well. It would also be extremely valuable to add structure information to the data, which may be able to give more actionable insights for materials design, as structure is very closely linked to property. Finally, the analytics framework developed and used in this paper can be used for building prediction models for other desired target properties, such as % Elongation, the data for which is also available in the NIMS dataset.

AVAILABILITY OF SUPPORTING DATA

The raw data used in this study was obtained from the publicly available NIMS MatNavi dataset, and was preprocessed as described in the paper. The preprocessed data is available as Additional file 1 of this paper.

ACKNOWLEDGEMENTS

Tbe authors are grateful to NIMS to make the raw data on fatigue steel strength publicly available. This work is supported in part by the following grants: NSF awards CCF-0833131, CNS-0830927, IIS-0905205, CCF-0938000, CCF-1029166, ACI-1144061, and IIS-1343639; DOE awards DE-

FG02-08ER25848, DE-SC0001283, DE-SC0005309, DESC0005340, and DESC0007456; AFOSR award FA9550-12-1-0458.

COMPETING INTERESTS

The authors declare that they have no competing interests.

AUTHORS' CONTRIBUTIONS

AA, PDD, AC applied one or more of the data analytics techniques used in this work. ANC provided data mining guidance and expertise; GPB and SRK provided materials science guidance and expertise. AA led the writing of the paper with inputs from all other authors. All authors read and approved the final submitted manuscript.

REFERENCES

1. Committee on Integrated Computational Materials Engineering N. R. C.: Integrated Computational Materials Engineering: A Transformational Discipline for Improved Competitiveness and National Security. 2008.http://www.nap.edu/openbook.php?record_id=12199
2. National Science and Technology Council: Materials genome initiative for global competitiveness. Technical report, National Science and Technology Council. 2011.http://www.whitehouse.gov/sites/default/files/microsites/ostp/materials_genome_initiative-final.pdf
3. Kalidindi SR, Niezgoda SR, Salem AA: Microstructure informatics using higher-order statistics and efficient data-mining protocols. *JOM - J Minerals, Met Mater Soc* 2011, 63(4):40–41.
4. Rajan K: Materials informatics. *Materials Today* 2005, 8(10):38–45.
5. Hey T, Tansley S, Tolle K: The Fourth Paradigm: Data-Intensive Scientific Discovery. Microsoft Research, 1st edition. 2009.http://research.microsoft.com/en-us/collaboration/fourthparadigm/ ISBN: 0982544200, URL: .
6. Linden G, Smith B, York J: Amazon.com recommendations: item-to-item collaborative filtering. *Internet Comput IEEE* 2003, 7(1):76–80.

7. Mobasher B: Data mining for web personalization. In *Brusilovsky P, Kobsa A, Nejdl W (eds) The adaptive web. Lecture Notes in Computer Science, vol. 4321*. Berlin, Heidelberg: Springer-Verlag; 2007:90–135.

8. Zhou Y, Wilkinson D, Schreiber R, Pan R: Large-scale parallel collaborative filtering for the netflix prize. In*Proceedings of the 4th International Conference on Algorithmic Aspects in Information and Management*. Berlin, Heidelberg: AAIM '08, Springer; 2008:337–348.

9. Das AS, Datar M, Garg A, Rajaram S: Google news personalization: Scalable online collaborative filtering. In*Proceedings of the 16th International Conference on World Wide Web*. New York, NY, USA: WWW '07, ACM; 2007:271–280.

10. *URL: Walmart is making big data part of its DNA*. Bigdata startups, 2013, . http://www.bigdata-startups.com/BigData-startup/walmart-making-big-data-part-dna/

11. King M: URL: Data Mining the TARGET way. 2012.http://www.slideshare.net/ipullrank/datamining-the-target-way

12. Rajan K, Suh C, Mendez P: Principal component analysis and dimensional analysis as materials informatics tools to reduce dimensionality in materials science and engineering. *Stat Anal Data Min* 2009, 1: 361–371.

13. Suh C, Rajan K: Virtual screening and qsar formulations for crystal chemistry. *QSAR & Comb. Sci* 2005, 24(1):114–119.

14. Nowers JR, Broderick SR, Rajan K, Narasimhan B: Combinatorial methods and informatics provide insight into physical properties and structure relationships during ipn formation. *Macromol Rapid Commun* 2007, 28: 972–976.

15. Gadzuric S, Suh C, Gaune-Escard M, Rajan K: Extracting information from the molten salt database. *Metallogr Mater Trans A* 2006, 37(12):3411–3414.

16. George L, Hrubiak R, Rajan K, Saxena SK: Principal component analysis on properties of binary and ternary hydrides and a comparison of metal versus metal hydride properties. *J Alloys Compounds* 2009, 478(1–2):731–735.

17. Singh S, Bhadeshia H, MacKay D, Carey H, Martin I: Neural network analysis of steel plate processing. *Iron-mak Steelmak* 1998, 25: 355–365.

18. Fujii H, MacKay D, Bhadeshia H: Bayesian neural network analysis of fatigue crack growth rate in nickel base superalloys. *ISIJ INT* 1996, 36: 1373–1382. 10.2355/isijinternational.36.1373

19. Hancheng Q, Bocai X, Shangzheng L, Fagen W: Fuzzy neural network modeling of material properties. *J Mater Process Technol* 2002, 122(2–3):196–200.

20. Gopalakrishnan K, Ceylan H, Kim S, Khaitan SK: Natural selection of asphalt mix stiffness predictive models with genetic programming. *ANNIE Int Eng Syst Artif Neural Netw* 2010, 20: 10.

21. Gopalakrishnan K, Manik A, Khaitan SK: Runway stiffness evaluation using an artificial neural systems approach. *Int J Electrical Comput Eng* 2006, 1(7):496–502.

22. Wen YF, Cai CZ, Liu XH, Pei JF, Zhu XJ, Xiao TT: Corrosion rate prediction of 3c steel under different seawater environment by using support vector regression. *Corrosion Sci* 2009, 51(2):349–355.

23. Rao BV, Gopalakrishna SJ: Hardgrove grindability index prediction using support vector regression. *Int J Miner Process* 2009, 91(1–2):55–59.

24. Gautham BP, Kumar R, Bothra S, Mohapatra G, Kulkarni N, Padmanabhan KA: More Efficient ICME through Materials Informatics and Process Modeling. In *Proceedings of the 1st World Congress on Integrated Computational Materials Engineering (ICME) (eds J. Allison, P. Collins and G. Spanos)*. Hoboken, NJ, USA: John Wiley & Sons, Inc.; 2011.

25. Dieter GE: Mechanical Metallurgy. Mc Graw-Hill Book Co. 3rd edition, ISBN: 0–07–016893–8 26. Deshpande PD, Gautham BP, Cecen A, Kalidindi S, Agrawal A, Choudhary (2013) Application of Statistical and Machine Learning Techniques for Correlating Properties to Composition and Manufacturing Processes of Steels. In *2nd World Congress on Integrated Computational Materials Engineering*. John Wiley & Sons, Inc.; 1986:155–160. ISBN: 9781118767061

26. Deshpande PD, Gautham BP, Cecen A, Kalidindi S, Agrawal A, Choudhary: Application of Statistical and Machine Learning Techniques for Correlating Properties to Composition and Manufacturing Processes of Steels. In *2nd World Congress on Integrated Computational Materials Engineering*. John Wiley & Sons, Inc.; 2013:155–160. ISBN: 9781118767061

27. URL: National Institute of Materials Science . http://smds.nims.go.jp/fatigue/index_en.html .

28. Weher E, Allen EL: An introduction to linear regression and correlation. (a series of books in psychology.) w. h. freeman and comp., San Francisco 1976. 213 s., tafelanh., s 7.00. *Biom J* 1977, 19(1):83–84.

29. Wang Y: A new approach to fitting linear models in high dimensional spaces. 2000.http://books.google.com/books?id=Z0OntgAACAAJ

30. Wang Y, Witten IH: Modeling for optimal probability prediction. In *Proceedings of the Nineteenth International Conference on Machine Learning*. Morgan Kaufmann Publishers; 2002:650–657.

31. URL: Robust Fit Regression, Mathworks . http://www.mathworks.in/help/stats/robustfit.html .

32. Aha DW, Kibler D: *Instance-based learning algorithms. Machine Learning Vol. 6*. Boston,: Kluwer Academic Publishers; 1991. pp 37–66 pp 37–66

33. Cleary JG, Trigg LE: K*: An instance-based learner using an entropic distance measure. In *Proceedings of the 12th International Conference on Machine Learning*. Morgan Kaufmann Publishers; 1995:108–114.

34. Kohavi R: The power of decision tables. In *Proceedings of the 8th European Conference on Machine Learning*. London, UK: ECML '95, Springer-Verlag; 1995:174–189.

35. Vapnik VN: *The nature of statistical learning theory. Information Science and Statistics Series*. New York, NY, USA: Springer-Verlag New York, Inc.; 1995. ISBN: 0-387-94559-8

36. Bishop C: *Neural Networks for Pattern Recognition*. USA: Oxford University Press; 1995. ISBN: 0198538642

37. Fausett L: *Fundamentals of Neural Networks*. New York: Prentice Hall, Pearson; 1994. ISBN: 0133341860

38. Witten IH, Frank E: *Data Mining: Practical Machine Learning Tools and Techniques. The Morgan Kaufmann Series in Data Management Systems,*. Morgan Kaufmann Publishers; 2005. ISBN: 0120884070

39. Wang Y, Witten IH: Induction of model trees for predicting continuous classes. In *Proc European Conference on Machine Learning Poster Papers*. Prague, Czech Republic; 1997:128–137.

40. Quinlan JR: Learning with continuous classes. In *5th Australian Joint Conference on Artificial Intelligence*. World Scientific; 1992:343–348.

41. Breiman L, Friedman J, Olshen R, Stone C: *Classification and Regression Trees*. Monterey, CA: Wadsworth and Brooks; 1984.

42. R Development Core Team: *R: A Language and Environment for Statistical Computing*. R Foundation for Statistical Computing, Vienna, Austria: R Foundation for Statistical Computing; 2011. ISBN 3-900051-07-0. .http://www.R-project.org.

43. MATLAB: *Version 7.10.0 (R2010a)*. Natick, Massachusetts: The MathWorks Inc.; 2010.

44. Hall M, Frank E, Holmes G, Pfahringer B, Reutemann P, Witten IH: The weka data mining software: An update.*SIGKDD Explorations Newsletter* 2009, 11(1):10–18.

10
CHAPTER

Multivariate Analysis for Fourier Transform Infrared Spectra of Complex Biological Systems and Processes

Diletta Ami[1, 2], Silvia Maria Doglia[1, 2] and Paolo Mereghetti [3]

[1] Department of Biotechnology and Biosciences, University of Milano-Bicocca, Milano, Italy

[2] Consorzio Nazionale Interuniversitario per le Scienze Fisiche della Materia (CNISM), UdR Milano-Bicocca, Milan, Italy

[3] Center for Nanotechnology Innovation @NEST, Italian Institute of Technology (IIT), Pisa, Italy

1. INTRODUCTION

Fourier transform infrared (FTIR) spectroscopy is a label-free and non invasive technique that exerts an enormous attraction in biology and medicine, since it allows to obtain in a rapid way a biochemical fingerprint of the sample under investigation, giving information on its main biomolecule content. This spectroscopic tool is successfully applied not only to the study of the structural properties of isolated biomolecules, such as proteins, nucleic acids, lipids, and carbohydrates, but also to the characterization of complex biological systems, for instance intact cells, tissues, and whole model organisms.

In particular, FTIR microspectroscopy, obtained by the coupling of an infrared microscope to a FTIR spectrometer, makes it possible to collect the IR spectrum from a selected sample area down to ~ 20 microns x 20 microns when conventional IR source and detector are employed, and down to of a few micrometers when more specialized and sensitive detectors and the highly brilliant synchrotron light source are used. In this way, FTIR microspectroscopy provides detailed information on several biological processes in situ, among which stem cell differentiation [1-5], somatic cell reprogramming [6], cell maturation [7, 8], amyloid aggregation

[9-12] and cancer onset and progression [13-15], making it possible to disclose the infrared response not only from single cells, but also from subcellular compartments [8, 16, 17].

The FTIR spectra of biological systems are very complex since they consist of the overlapping absorption of the main biomolecules; for this reason, to pull out the significant and non-redundant information contained in the spectra it is necessary to apply an appropriate multivariate analysis, able to process very high-dimensional data. This is even more crucial when time-dependent biological processes, such as cell maturation or differentiation, are studied. Indeed, in this case it is fundamental to be able to extract from the spectral data the relevant information of the process you are investigating [18-21].

In Figure 1 we schematized the procedure that should be followed to successfully tackle the FTIR characterization of complex biological systems.

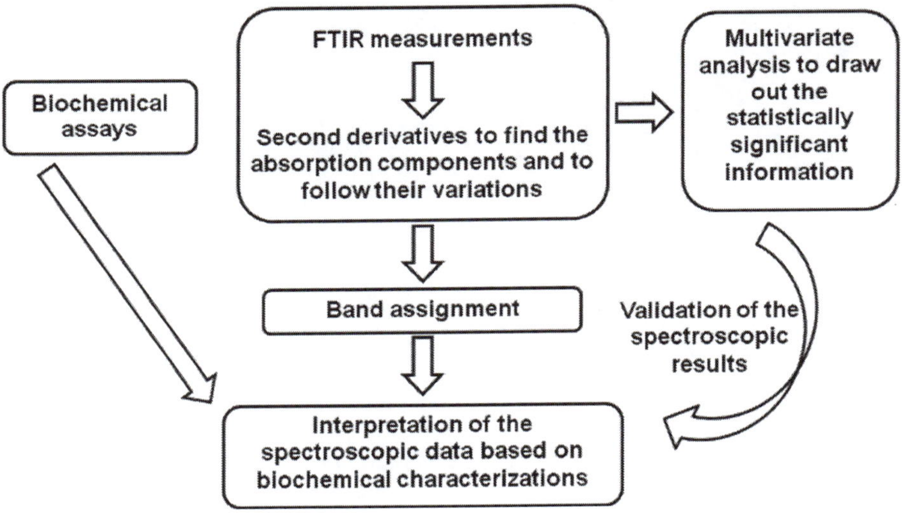

Figure 1. Scheme of the FTIR approach to study complex biological systems. The IR absorption spectra are analysed by resolution enhancement approaches (e.g. second derivatives) to resolve the overlapped absorption components and to monitor their variations during the process under investigation. The spectroscopic results are validated by an appropriate multivariate analysis approach, to identify firstly specific marker bands of the studied process. The interpretation of the spectroscopic data should be then confirmed by standard biochemical assays.

Several multivariate analysis approaches exist and for the scope of this book they can be divided into two main categories: regression and classification techniques. In the first category fall all methods that allow to derive a model describing the relationship between two sets of variables. The second category includes techniques to split observations into groups or classes.

In this chapter, we will firstly introduce the most widely used multivariate analysis approaches in the field of spectroscopy.

We will then illustrate the basic principles and experimental details for the application of principal component - linear discriminant analysis (PCA-LDA) to the analysis of FTIR spectral data of complex biological systems. The potential of these combined tools will be described on illustrative examples of cell biological process studies. In particular, we will discuss in details its application on our FTIR study of murine oocytes characterized by two different types of chromatin organisation around the nucleolus, strongly affecting their development after fertilization. In this case, PCA-LDA analysis made it possible to identify not only the maturation stage in which the fate separation between the two kinds of oocytes occurred, but also to disclose the most significant cellular processes responsible for the different oocyte destiny, thus validating the visual inspection of the infrared spectra [7].

2. FTIR MICROSPECTROSCOPY OF COMPLEX BIOLOGICAL SYSTEMS

Fourier transform infrared (FTIR) microspectroscopy is a powerful technique that allows to obtain a molecular fingerprint of the sample under investigation in a rapid and non-invasive way. In the case of complex biological systems it provides simultaneously, in a single measurement, information on the main biomolecules, such as lipids, proteins, nucleic acids, and carbohydrates, requiring also a very limited amount of sample. For these reasons, it became recently a very attracting tool for biomedical research [20, 22-24], being successfully employed for the study of several biological systems, from intact cells [6, 7, 25] to tissues [11, 26, 27] and whole model organisms (i.e. the nematode*Caenorhabditis elegans*) [9, 28].

As an example, in Figure 2 it is reported the FTIR absorption spectrum of a single intact murine oocyte. As shown, its IR response is very complex, being due to the absorption of the main biomolecules. In particular, between 3050 - 2800 cm^{-1} and 1500 - 1350 cm^{-1} the absorption of the lipid acyl chains occurs, while around 1740 cm^{-1} the ester carbonyl absorbs [29]. Moreover, the amide I and amide II bands - mainly due to the C=O stretching and the NH bending of the peptide bond respectively - give information on the protein secondary structure [30], while the spectral range between 1000 and 800 cm^{-1} is very informative on nucleic acid absorption, since it is due in particular to sugar vibrations sensitive to their conformation and to backbone vibrational modes [31, 32]. Finally, we should also mention the very complex spectral range between 1250 - 1000 cm^{-1}, mainly due to phosphodiester groups of nucleic acids and phospholipids and to the C-O absorption of glycogen and other carbohydrates [31, 33, 34].

Making it possible to obtain a sample biochemical fingerprint in a rapid and non destructive way, FTIR microspectroscopy is widely applied to the in situ characterization of cellular processes, such as cell maturation, differentiation, and reprogramming [3, 5-7, 25, 35], and to the detection of several diseases, as, for instance, cancer [13-15] and neurodegenerative disorders [10, 11], whose onset is accompanied by changes in the composition and structure of several biomolecules.

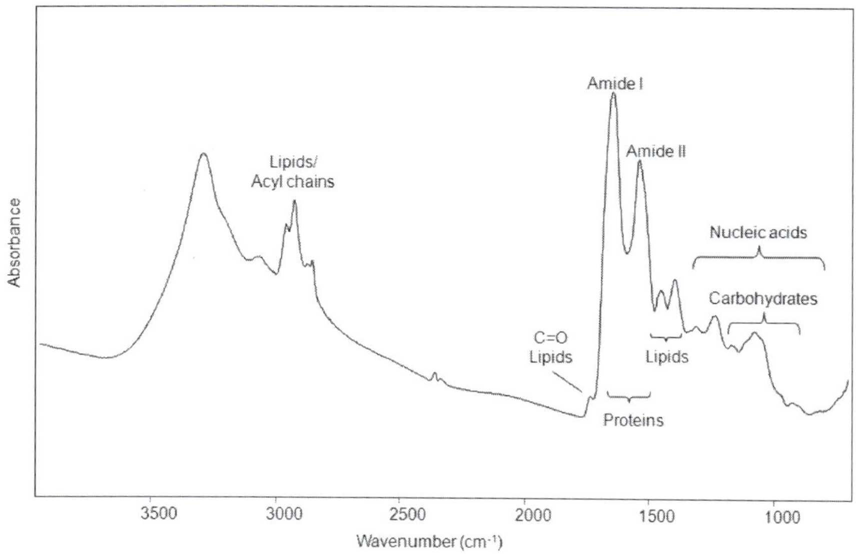

Figure 2. FTIR absorption spectrum of a single intact murine oocyte. The measured absorption spectrum of a single intact murine oocyte (surrounded nucleolus, MI 10 H) is reported without any corrections. The oocyte - deposited on a BaF_2 window - was measured in transmission by the IR microscope UMA 500, coupled to the FTIR spectrometer FTS 40A (both from Digilab), at a resolution of 2 cm^{-1}. The absorption regions of the main biomolecules are indicated.

Since water has a strong absorption in the mid-infrared spectral range, samples have to be dried rapidly before IR measurements, in particular when working in transmission mode (see for details the following paragraph). The suitability of such "dry-fixing" has been proved by Raman spectroscopy, a vibrational tool complementary to FTIR, whose response is not affected by water. In particular, Raman measurements performed on differentiating human embryonic stem cells, hydrated and dry-fixed, demonstrated that the rapid desiccation didn't affect the spectroscopic response of the main biomolecules. Indeed, in both cases the same temporal pattern of the differentiation marker bands - due to tryptophan, nucleic acid backbone and base vibrations - was observed during the biological process under investigation [36].

We should add that to obtain reliable results on the studied process it is crucial to standardize firstly the sample preparation, since - for instance - metabolic changes due to cell aging could result in significant spectral changes that could, in turn, hide the IR response specifically due to the process of interest, as it has been recently reported in the literature [37]. For these reasons, it is fundamental to check accurately the stage of cell growth in culture before performing spectroscopic measurements.

We should also briefly mention that, before spectral analyses, the measured IR spectra could require some corrections due to artifacts that can interfere with the spectroscopic response. For instance, single cells, or subcellular compartments, or particles of the size of the same order of that of the incident infrared light (~3-10 microns) could give rise to Mie scattering, that significantly distorts the measured spectrum, causing misinterpretation of the results. For this reason, before further analyses, it is strongly recommended to correct the measured spectra with opportune algorithms specifically developed to this aim [38].

Since the IR spectra of complex biological systems are due to the overlapping spectral features of multiple components, their analysis requires often the employment of resolution enhancement procedures to better resolve their absorption bands, an essential prerequisite for the identification of peak positions and their assignment to the vibrational modes of the different molecules. Among these, second derivative analysis is widely applied, as described in [39]. Since second derivative band intensity is inversely proportional to the square of the original band half-width, this procedure introduces an enhancement of sharp lines, as those due to vapour and noise. For this reason, this analysis requires spectral data free of vapour absorption and with excellent signal to noise ratio.

Furthermore, due to the intrinsic complexity of biological systems, their spectral analysis requires the support of appropriate multivariate analysis approaches able to tackle the study of high-dimensional data, to verify firstly the reproducibility of the results and then to extract the most significant spectral information [18-21] (see for details paragraph 4).

3. FTIR MICROSPECTROSCOPY: TECHNICAL CONSIDERATIONS

FTIR microspectroscopy is realized coupling to a FTIR spectrometer an infrared microscope characterized by an all reflecting optics, since typical lenses and condensers of visible microscopy - being made of glass, not transparent to the IR radiation - cannot be employed.

The main advantage of FTIR microspectroscopy is that it offers the possibility to study selected areas of the sample under investigation, resulting particularly useful in the case of systems characterized by an intrinsic heterogeneity, such as biological systems.

Two main types of IR microscopy exist, depending on the detector employed, and both equipped with an IR thermal source (globar), whose spatial resolution is diffraction-limited.

The first, conventional, generally equipped with a nitrogen cooled mercury cadmium telluride (MCT) detector, makes it possible to measure IR absorption spectra from a microvolume within the sample, selected by a variable aperture of the microscope, whose side can be adjusted down to a few tens of microns.

The second type of IR microscope, more advanced, is equipped with a focal plane array (FPA), consisting of an array of infrared detector elements, that enables not only to collect the IR absorption spectrum of the sample, but also an IR chemical imaging, where the image contrast is given by the response of selected sample regions to particular IR wavenumbers. Depending mainly on the detection array, the spatial resolution in this kind of microscopy is approximately between 20 and 5 microns, making it possible to reach, therefore, a resolution near to the diffraction limit.

We should, however, add that the use of a synchrotron IR light source, with a brightness of at least two orders of magnitude higher than that of a conventional thermal source, makes it possible to achieve diffraction-limited spatial resolution with enhanced signal-to-noise ratio. In this way, synchrotron light could allow to explore the IR spectra at the subcellular level.

A final remark should be done concerning the spectral acquisition mode. Indeed, infrared measurements can be mainly performed in transmission, reflectance or attenuated total reflection (ATR) mode. Typically, measurements on complex biological systems are performed in transmission mode, using appropriate IR transparent supports for the deposition of the sample, such as BaF_2, CaF_2, ZnSe. In this case, the IR beam goes through the sample, that - depending mainly on its molar extinction coefficient - should have a uniform thickness, not exceeding 15-20 microns.

Moreover, in reflectance mode - where the sample is placed onto proper reflective slides - the IR beam passes the sample, is reflected by the slide, and passes the sample again. In particular, the sample slides reflect mid-infrared radiation almost completely and usually are also transparent to visible light, allowing sample inspection by a conventional light microscope. This approach is, for instance, useful for tissue characterizations.

Finally, in the ATR approach, where the sample is placed into contact with a higher refractive index and an IR transparent element (mainly germanium and diamond), samples with higher thickness than in transmission can be processed. In particular, the IR beam reaches the interface between the ATR support and the sample at an angle larger than that corresponding to the total reflection. In this way the beam is totally reflected by the interface and penetrates into the sample as an evanescent wave, where it can be absorbed. The beam penetration depth is of the order of the IR wavelength (a few micrometers) and depends on the wavelength, the incident angle, as well as on the refractive indices of the sample and of the ATR element. Furthermore, it should be noted that this kind of approach makes it possible to measure also samples not necessarily deposited onto an IR transparent support, as in ATR measurements it is only required that the sample be in close contact with the ATR element.

For a review of the technical aspects of FTIR microspectroscopy, see [40-42].

4. MULTIVARIATE ANALYSES

4.1. Introduction To Multivariate Analysis

Several phenomena can only be described or explained by taking into account several variables at the same time. These cases represent the realm of the Multivariate statistical analysis (MVA).

We now define the structure of our data that will be kept throughout the text for all described techniques. For a given phenomenon we perform a certain measurement and store the value in a uni- or multivariate variable called

$$\mathbf{y} = (y_1, y_2, \ldots, y_m)^T,$$

where m is the number of independent variables. The same measurement can be repeated several times on the same sample or on different samples. We then define a group as a collection of two or more replica of the same experiment and we also define the term instance or observation to refer to a specific experiment within one group.

Each instance associated to the variable y is stored in a matrix **Y** composed of *n* rows (the observations) and *m* columns (the independent variables).

$$\mathbf{Y} = \begin{pmatrix} y_{11} & y_{12} & \cdots & y_{1m} \\ \vdots & \vdots & \vdots & \vdots \\ y_{n1} & y_{n1} & \cdots & y_{nm} \end{pmatrix} = (\mathbf{y}_1, \mathbf{y}_2, \ldots, \mathbf{y}_n)^T \tag{1}$$

Each element of matrix **Y** can be indicated as

$y_{ij}y_{ij}$

where *i* indicate the observation and *j* is an independent variable. In some cases we want to find or explain the relationship between the independent variables **Y** and another set of uni- or multivariate variables **Z**. Similarly to the **Y** matrix, the matrix **Z** has *n* rows, one for each observation and *m* columns, the dependent variables.

$$\mathbf{Z} = \begin{pmatrix} z_{11} & z_{12} & \cdots & z_{1m} \\ \vdots & \vdots & \vdots & \vdots \\ z_{n1} & z_{n1} & \cdots & z_{nm} \end{pmatrix} = (\mathbf{z}_1, \mathbf{z}_2, \ldots, \mathbf{z}_n)^T \tag{2}$$

The matrix **Y** (composed of the independent variables y) represents the only input for several multivariate techniques described here; in some other cases the matrices **Y** and **Z** (composed of the dependent variables z) are both required.

In the following part, we will make a distinction between regression and classification techniques. However, it should be clear that the separation between these two domains is not always sharp and the same technique can be either used for regression or for classification purposes.

4.2. Multivariate Regression Techniques

4.2.1. Linear multivariate regression (lmvr)

LMVR (or MLR) can be used to model linear relationships between one or more **z** (dependent variable) and one or more y (independent variable). In the most general case, we have *n* independent multivariate variables y represented by the matrix **Y** and the corresponding response multivariate variable **z**, stored in the matrix **Z**.

The LMVR is based, as many other statistical techniques, on the generalized linear model: $\mathbf{Z} = \beta\mathbf{Y} + \varepsilon$ where β is a matrix containing the parameters to be estimated, and ε is a matrix which models the errors or noise. The coefficients β are usually estimated using the ordinary least square, which consists of minimizing the sum of the square differences of the n observed y's from their modeled values. Mathematically, the optimal values of β are obtained by $\beta = \left(\mathbf{Y}^T\mathbf{Y}\right)^{-1}\mathbf{Y}^T\mathbf{Z}$. To apply the least square method we must have $n - 1 > m$ (e.g. the number of observation must be larger than the number of variables, which is often not the case), otherwise the matrix $\mathbf{Y}^T\mathbf{Y}$ is singular and cannot be inverted. Another common problem is the correlation between variables; more specifically, none of the independent variables must be a linear combination of any other. This phenomenon is called "multicollinearity" [43, 44] and it will be explained in more details in section 4.2.3.

4.2.2. Non-linear multivariate regression (nlmvr)

In some cases linear models cannot be used and one could try to apply non-linear models.

Common models which frequently apply to natural phenomena are the exponentials (which, indeed, is a transformed linear model. A linear model can be applied upon on the logarithm of the data), logistic models or power law models.The regressed model has the general form of $\mathbf{Z} = \beta f(\mathbf{Y}) + \varepsilon$, where $f(\mathbf{Y})$ can be any non-linear function. The optimal values for the coefficients β can be obtained using deterministic optimization algorithm such as the conjugate gradients [45] or the Levenberg-Marquard method [46, 47], or stochastic algorithm such as genetic algorithms [48].

4.2.3. The multicollinearity problem

When the number of observations is smaller than the number of variables (as it often happens for spectral data), the matrix $\mathbf{Y}^T\mathbf{Y}$ is singular and is not invertible. This rules out the possibility of using standard linear multivariate techniques (LMVR) based on the least square criterion, as the solution will not be unique.

Increasing the number of observations (above the number of variables) will not always solve the problem. This is due to the so-called near-multicollinearity which means that some variables can be written approximately as linear functions of other variables. This problem is often found among spectral measurements. Even if the solution will be mathematically unique, it may be unstable and lead to poor prediction performances.

Linearly correlated or quasi-linearly correlated variables have to be removed prior to apply a regression method. In the following sections, we will describe two methods that are frequently used to remove correlations among variables, namely principal component analysis (PCA) and partial least squares (PLS).

4.2.3.1. Principal Component Analysis (PCA)

We should first recall the structure of the data. Suppose that we have *n* observations, each one defined by a vector yi composed of *m* variables, where *i=1,2,...,n* stands for the i-th observation. The matrix of the original data **Y** is then composed by *n* rows (the observations) and *m* columns (the variables).

By using PCA, our intent is to develop a smaller number of uncorrelated artificial variables, called principal components (PC), that will account for most of the variance in the observed variables. The new uncorrelated variables are obtained as linear combination of the original data as T=AY . Correlation among variables can be measured using the covariance matrix.

Given the sample mean of the m-dimensional vector yi,

$\mathbf{y}_i$, $\langle \mathbf{y} \rangle = \frac{1}{n}\sum_{i=1}^{n} \mathbf{y}_i$, an unbiased estimator of the sample covariance matrix is $\mathbf{S} = \frac{1}{n-1}\sum_{i=1}^{n}(\mathbf{y}_i - \langle \mathbf{y} \rangle)(\mathbf{y}_i - \langle \mathbf{y} \rangle)^T$

For uncorrelated variables, the off-diagonal values of the sample covariance matrix are zero, that is, S is diagonal. The covariance of linearly transformed variables T=AY is equal to $\mathbf{S}_T = \mathbf{ASA}^T$, where **S** is the sample covariance of the original data **Y** [49].

Thus, we want to find the matrix **A** such that the covariance matrix of the transformed data, ST , is diagonal, which corresponds to find the eigenvectors of the covariance matrix and the corresponding eigenvalues. The eigenvalues, which coincide with the matrix ST , are the sample variance of the principal components **T** and are ranked according to their magnitude. The first principal component is then the linear combination with maximal variance (the largest eigenvalue). The second principal component is the linear combination with the maximal variance along a direction orthogonal to the first component, and so on [44].

The number of eigenvalues is equal to the number of original variables; however, since the eigenvalues are equal to the variance of the principal components and they are sorted in a decreasing order, the first*k* eigenvalues can account for a large portion of the variance of the data.

Hence, to describe our original dataset we can use only the first *k* uncorrelated principal components, instead of the complete set of redundant *m* variables. In matrix notation this can be written as

$T_k = A_k Y$, where A_k is the eigenvector matrix truncated to the k-th eigenvector, and T_k is the matrix of the first *k* principal components, also called score matrix [50].

Choosing which and the number of principal components that should be retained in order to summarize our data is a task that can be solved using several strategies [43, 49]. For example, one way commonly used is to retain the first k principal components that explain a given total percentage of the variance, e.g. 90% [43, 44]. Another rule is to plot the eigenvalues in decreasing order. Moving from left to right, the eigenvalues usually have an initial steep drop followed by a slow decrease. All the components after the elbow between the steep and the flat part of the curve should be discarded. This test is called screen plot.

Alternatively, one can select the principal components that can be associated to a physical meaning related to the studied system. For example, following the differentiations of a cell line growing in different experimental conditions, one principal component may represents the different conditions, while another PC may describe the maturation stage of the cells. None of the above methods are better than the other; usually more than one test should be done and the results compared.

The principal component analysis allows to obtain uncorrelated variables and then to remove the multicollinearity problem.

4.2.3.2. Principal Component Regression (PCR): multivariate regression following PCA

Once a set of *k* principal components has been obtained using the PCA method, they can be used as input variables for a multivariate regression analysis instead of the original data. The regression equation

$Z=\beta Y+\varepsilon Z=\beta Y+\varepsilon$

, shown in section 4.2.1, can be written as

$Z=\beta Tk+\varepsilon Z=\beta Tk+\varepsilon$

, where $\mathbf{T}_k$ is the matrix of the principal components (scores matrix) and the regression coefficients

$\beta\beta$

can be estimated by least squares. When the number of principal components is equal to the number of variables, this method becomes equivalent to the LMVR. By removing correlations in the original data, the PCR method allows to perform linear regression on a multicollinear dataset.

4.2.3.3. Partial Least Squares (PLS)

Another way to face the multicollinearity problem is to use PLS. The goal of PLS regression is to predict **Z** from **Y** and to describe their common structure [50].

In the PCR method described above, the principal components are selected based on their ability of explaining the variance of the **Y** matrix (the dependent variable matrix). By contrast, PLS regression finds components from **Y** that are also relevant for **Z**. Specifically, PLS regression searches for a set of components that performs a simultaneous decomposition of **Y** and **Z**, with the constraint that these components explain as much as possible the covariance between **Y** and **Z**. In this way, compared to the PCR, the principal components contain more information about the relationship between predictors and dependent variables [50]. For categorical dependent variables, the PLS method takes the name of partial least square discriminant analysis (PLS-DA) [43].

4.3. Multivariate Classification Techniques

Classification methods can be divided into two main categories, supervised and unsupervised. Supervised techniques require the knowledge of the group membership of the observations and can be used to understand the structure of the data, e.g. why certain observations belong to a given group. Moreover, once the classification model is calibrated on a "training" dataset, it can be used in a predictive way to group observations whose group membership is unknown.

On the other hand, unsupervised methods try to group the observations without any knowledge of the group membership.

In the following paragraph, we will describe the main multivariate classification approaches.

4.3.1. Discriminant analysis (da)

Discriminant analysis is mainly a supervised technique which was originally developed by Ronald Fisher as a way to subdivide a set of taxonomic observations into two groups based on some measured features [51]. Later, DA was extended to treat cases where there are more than two groups, the so-called "multiclass discriminant analysis" [49, 52, 53].

DA can have mainly two objectives. First, it can be used in a supervised way to describe and explain the differences among the groups. As we will see later, mathematically DA finds the optimal hyperplane that separates the groups among each other. Or, in other words, it finds the optimal linear combination of the original variables that maximizes the distance among the groups. The transformed observations are called discriminant functions.

The use of a linear combination implies that each original variable is weighted by a coefficient which can be used to study the relative importance of the variable in the separation among the groups. A second possible role of DA is to classify observations into groups. An observation, which has to be assigned to a group, is evaluated by a discriminant function (already calibrated on another dataset) and it is assigned to one of the groups at which most likely it belongs [43, 44, 49]; in this view DA is used as an unsupervised method.

When only linear transformations are applied to the variables used as DA input, the discriminant analysis is called linear discriminant analysis (LDA).

In some cases, LDA alone is not suitable and the original variables can be mapped to a new space via any non-linear function. Then, the LDA is applied in this non-linear space (which is equivalent to non-linear classification in the original space). This procedure can be seen under several names such as "non-linear DA" (NLDA) or "kernel Fisher discriminant analysis" (KFD) or "generalized discriminant analysis".

In the following sections we will focus on LDA, first describing the descriptive approach and subsequently the classification approach.

4.3.1.1. Linear DA (LDA) as a descriptive method

The initial dataset is an ensemble of multivariate observations partitioned into *G* distinct groups (e.g. different experimental treatments, times or conditions). Each of the *G* groups contains n_g observations, where *g* runs from *1* to *G* and refers to the g-th group. The multivariate observation vectors can be written as y_{gj} where *g* is the g-th group and *j* is the j-th observation. The vector has size*m*, which corresponds to the number of variables.

Our goal in LDA is to search for the linear combination that optimally separates our multivariate observation into *G* groups. The linear transformation of y_{gj} is written as

$$z_{gj} = \mathbf{w}^T \mathbf{y}_{gj} \tag{3}$$

Since z_{gj} is a linear transformation of z_{gj}, the mean of the group *g* of the transformed data can be written as

$$\langle z_g \rangle = \mathbf{w}^T \langle \mathbf{y}_g \rangle \tag{4}$$

where z_{gj} is the mean, of the observations within a group, obtained as

$$\langle \mathbf{y}_g \rangle = \sum_{j=1}^{n_g} \mathbf{y}_{gj} / n_g$$

We now introduce the between groups sum of squares **B** in equation 5 (measure of the dispersion among the groups) and the within-group sum of squares **E** in equation 6 (measure of the dispersion within one group). First, we define them for the uni-dimensional case relatively to the untransformed data:

$$\mathbf{B}(y) = \sum_{g=1}^{G} \left(\langle y_g \rangle - \langle y \rangle \right)^2 \tag{5}$$

and

$$\mathbf{E}(y) = \sum_{g=1}^{G} \sum_{j=1}^{n_g} \left(y_{gj} - \langle y_g \rangle \right)^2 \tag{6}$$

where $\langle \mathbf{y} \rangle = \frac{1}{G} \sum_{g=1}^{G} \frac{1}{n_g} \sum_{j=1}^{n_g} \mathbf{y}_{gj}$ is the total average of the data.

Analogously, in the multivariate case (where each observation is constituted by *m* variables) we have the two matrices:

$$\mathbf{B}(\mathbf{y}) = \sum_{g=1}^{G} n_g \left(\langle \mathbf{y}_g \rangle - \langle \mathbf{y} \rangle \right) \left(\langle \mathbf{y}_g \rangle - \langle \mathbf{y} \rangle \right)^T \tag{7}$$

and

$$\mathbf{E}(\mathbf{y}) = \sum_{g=1}^{G} n_g \sum_{j=1}^{n_g} \left(\langle \mathbf{y}_{gj} \rangle - \langle \mathbf{y}_g \rangle \right) \left(\langle \mathbf{y}_{gj} \rangle - \langle \mathbf{y}_g \rangle \right)^T \tag{8}$$

Finding the optimal linear combination that separates our multivariate observations into *k* groups means to find the vector *w* which maximizes the rate between the between-groups sum of squares over the within-groups sum of squares. Using the equations for the transformed data (equations 3 and 4) into the equations 7 and 8, we can write:

$$\lambda = \frac{\mathbf{w}^T \mathbf{B}(\mathbf{y}) \mathbf{w}}{\mathbf{w}^T \mathbf{E}(\mathbf{y}) \mathbf{w}} = \frac{\mathbf{B}(\mathbf{z})}{\mathbf{E}(\mathbf{z})} \tag{9}$$

We want to find *w* such that λ is maximized.

Equation 9 can be rewritten in the form $\mathbf{w}^T(\mathbf{Bw}-\lambda\mathbf{Ew})=0$; then we search for all the non trivial ($w^T=0$ is excluded) solutions of this equation and we choose the one which gives the maximum value of λ. This means to solve the eigenvalue problem Bw−λEw=0 that can be written in the usual form:

$$(\mathbf{A}-\lambda\mathbf{I})\mathbf{w}=0 \tag{10}$$

where $\mathbf{A}=\mathbf{E}^{-1}\mathbf{B}$.

The solutions of equation 10 are the eigenvalues λ1,λ2,...,λm associated to the eigenvectors w1,w2,...,wm . The solutions are ranked for the eigenvalues λ1>λ2>...>λm. Hence, the first eigenvalue λ1 corresponds to the maximum value of equation 9.

The discriminant functions are then obtained considering only the first *s* positive eigenvalues and multiplying the original data by the eigenvectors

$$\mathbf{z}_1=\mathbf{w}_1^T\mathbf{Y},\mathbf{z}_2=\mathbf{w}_2^T\mathbf{Y},\ldots,\mathbf{z}_s=\mathbf{w}_s^T\mathbf{Y}.$$

Discriminant functions are uncorrelated but not orthogonal since the matrix

$A=E_{-1}B$ is not symmetric.

In many cases the first two or three discriminant functions account for most of $\lambda_1+\lambda_2+...+\lambda s$. This allows to represent the multivariate observations as 2 or 3 dimensional points which can be plotted on a scatter plot. These plots are particularly helpful to visualize the separation of our observations into the different groups. Moreover, we can deduce, looking at the scatter plot, the meaning of a given discriminant function, i.e. we can associate the discriminant function to a given property of the analyzed system.

The weighting vectors w1,w2,...ws are called unstandardized discriminant function coefficients and give the weight associated to each variable on every discriminant function.

If the variables are on very different scales and with different variance, to assess the importance of each variable in the group separation the standardized discriminant functions can be used. The standardization is done by multiplying the unstandardized coefficients by the square root of the diagonal element of the within-group covariance matrix.

Another way to assess the variable importance is to look at the correlation between each variable and the discriminant function. These correlations are called structure or loading coefficients. However, it has been shown that these parameters are intrinsically univariate and they only show how a single variable contributes to the separation among groups, without taking into account the presence of the other variables [49].

4.3.1.2. Linear as a classification method

After a set of discriminant functions are calibrated as described in the previous section, the discriminant analysis can be applied to classify new observations into the most probable groups. From this point of view, the linear discriminant analysis becomes a predictive tool, since it is able to classify observations whose group membership is unknown [43, 49]. The discrimination ability of our LDA model can be tested by a procedure called "re-substitution" [49]. This method consists of producing an LDA model using our dataset (i.e. finding the optimal w). Then, each observation vector is re-submitted to the classification function (zgj=wTygj) and assigned to a group. Since we know the group membership of the submitted vector, we can count the number of observations correctly classified and the number of observations misclassified. To measure the classification accuracy we can count the number of observations correctly classified and the number of observations misclassified. Then, we can estimate the classification rate as the number of correctly classified observations over the total number of observations. In general, in evaluating the accuracy of a model, we have then to distinguish between two types of accuracy: the fitting accuracy and the prediction accuracy [43, 54].

The fitting accuracy is the ability to reproduce the data, namely how the model is able to reproduce the data that were used to build the model (the training set). This corresponds to the apparent classification rate and it is obtained using the re-substitution procedure.

The prediction accuracy is the ability to predict the value or the class of an observation, that was not included in the construction of the model. This kind of accuracy is often referred to as the ability of the model to generalize. The data used to measure this accuracy are called "test set". The prediction accuracy can be called "actual classification rate". This is mainly used in settings where the goal is prediction, and one wants to estimate how accurately a predictive model will perform in practice. To have an estimation of the actual classification rate, two main procedures can be applied: the hold-out and cross-validation [43].

In the hold-out, the dataset is divided into two partitions: one partition is used to develop the model (e.g. the discriminant functions) and the second partition is given as input to the model. The first partition is usually called "training set" or "calibration set", while the second partition is the validation set [54].

When the number of observations is small, the cross-validation is usually preferred over the hold-out. The basic idea of the cross-validation procedure is to divide the entire dataset into L disjoint sets. L-1 sets are used to develop the model (i.e. the calibration set on which the discriminant functions are computed) and the omitted portion is used to test the model (i.e. the validation set given as input to the model). This is repeated for all the L sets and an average result is obtained.

Apparent or actual classification accuracies can be summarized in a confusion matrix. As an example, total N observations, n_1, belong to the group 1 and

n2 belong to the group 2. C_{11} is the total number of observations correctly classified in group 1 and C_{12} is the total number of data misclassified in group 2. Similarly, C_{22} is the total number of observations correctly classified in group 2 and C_{21} is the number of misclassified in group 1.

The confusion matrix becomes then:

	Actual group	Predicted group
	1	2
1	C_{11}	C_{12}
2	C_{21}	C_{22}

and the accuracy (the actual or apparent classification rate (acr)) is computed as:

$$acr = \frac{C_{11} + C_{22}}{n_1 + n_2}$$

4.3.2. PCA-LDA

A powerful analysis tool is the combination of the principal component analysis with the linear discriminant analysis [52]. This is particularly helpful when the number of variables is large. In particular, if the number of observations (*N*) is less than the number of variables (*m*) - specifically $N-1<m$ - the covariance matrix is singular and cannot be inverted (see section 4.2.3.). We then need to find a way to reduce the number of variables, for example using the PCA [49, 55]. This procedure has been widely used for several problems in different fields [35, 52, 56-60]. The condition $N-1<m$ almost always appears in spectroscopy, where the number of observations (*N*) is usually 10^2 and the number of variables (*m*) is typically within 10^2 to 10^3.

Let's take into account the same situation described for the many group linear discriminant analysis. The original dataset is an ensemble of multivariate observations which is partitioned into *k* distinct groups. Again, we want to find the discriminant functions which optimally separate our multivariate observation into the *k* groups. Then, the discriminant functions can be used to identify the most important variables in terms of ability of distinguishing among the groups. Thus, first the original dataset is submitted to the PCA to reduce the number of variables; subsequently, the reduced dataset is analyzed using the LDA.

Another way that can be used instead of PCA is to perform the PLS.

4.3.3. PLS-LDA

In a way analogous to the PCA-LDA procedure, here we first apply the PLS algorithm to the original data and then the LDA on the selected principal components [61].

Given that the PLS searches for a set of components that performs a simultaneous decomposition of the dependent and independent datasets, the main difference with PCA-LDA is that the principal components resulting as output of PLS better describe the relationship between independent and dependent variables. This does not necessarily mean that this method is better in general. Indeed, applying PCA or PLS on the same dataset often leads to similar results [62, 63] and the classification accuracy or the descriptive ability is mostly determined by the underlying structure of the data which can make one of the two methods more suitable than the other.

4.3.4. CLUSTER ANALYSIS (CA)

The goal of cluster analysis is to find the best grouping of the multivariate observations such that the clusters are dissimilar to each other but the observations within a cluster are similar [44].

CA is an unsupervised technique, that is, the group membership of the observations (and often the number of groups) is not known in advance.

At first we have to define a measure of similarity or dissimilarity also called distance functions. The most common distance functions are: i) the Euclidean distance; ii) the Manatthan distance; iii) the Mahalanobis distance; iv) the maximum norm.

Based on the procedure they use, clustering algorithms can be divided into three main groups: hierarchical, partitional and density-based clustering. None of the following algorithms is better than the other. The choice of the clustering method strongly depends on the structure of the data and on which kind of results one would expect.

Hierarchical clustering algorithms can be again subdivided into agglomerative or divisive. The agglomerative clustering starts with all observations placed in different clusters and in each step an observation or a cluster of observations is merged into another cluster. The most commonly employed agglomerative clustering strategies are complete-linkage, average-linkage, single-linkage, centroid-linkage. The drawback of the agglomerative clustering algorithms is that observations cannot be moved among the clusters once a cluster is made.

The divisive method starts with one single cluster containing all observations and then it divides the cluster into two sub-clusters at each step. Divisive methods have the same drawback of the agglomerative clustering, that is, once a cluster is made, an observation cannot be moved to another cluster. Divisive methods are suited when large clusters are searched for.

The partitional algorithm assigns the observations to a set of clusters without using hierarchical approaches. One of the most used non-hierarchical approach is the k-means clustering.

The density-based clustering seeks to search for regions of high density without any assumption about the shape of the cluster.

4.4. Artificial Neural Networks (ANN)

The artificial neural networks are mathematical models that were developed in analogy to a network of biological neurons [64]. Mathematically, a neuron can be modeled as a switch that receives, as input, a series of values and produces an output consisting of a weighted sum of the input eventually transformed by a function f. Many neurons can be combined to create more complex networks. Depending on the type of neurons and on how the neurons are connected to each others, different kinds of neural networks can be created. The most common type of neural network is the feed-forward neural network, in which neurons are grouped into layers, each neuron of a layer is connected to all the neurons of the next layer and the information flows from the input to the output without loops. For a comprehensive description of neural networks and their applications see [54, 65].

5. APPLICATIONS OF MULTIVARIATE ANALYSIS TO SPECTROSCOPIC DATA OF COMPLEX BIOLOGICAL SYSTEMS

In the following, we will provide a few selected examples of the application of FTIR microspectroscopy coupled with multivariate analysis for biomedical relevant studies, with the aim to highlight the importance of linking the two approaches to extract the most significant spectral information from highly informative systems.

In some cases, PCA alone represents a powerful method for the analysis of multidimensional FTIR spectra. Indeed, several interesting works are reported in the literature, in which this approach is employed to support the spectroscopic investigation of complex biological systems and processes. For instance, synchrotron based FTIR microspectroscopy coupled with PCA has been applied to the characterization of human corneal stem cells [27, 66], in cancer research for the screening of cervical cancer [14], as well as to disclose the effects induced by a surface glycoprotein in colon carcinoma cells [67].

For instance, Matthew German and colleagues [68] coupled high-resolution synchrotron radiation-based FTIR (SR-FTIR) microspectroscopy with PCA to investigate the characteristics of putative adult stem cell (SC), transiently amplified (TA) cell, and terminally differentiated (TD) cell populations of the corneal epithelium. Using PCA, each spectrum, composed by many variables (the wavenumbers), is reduced to a point in a low dimensional space. Then, each observation can be visualized in a two or three dimensional score plot. Choosing the appropriate principal components, the authors were able to clearly distinguish the three cell populations confirming the ability of SR-FTIR microspectroscopy to identify SC, TA cell, and TD cell populations.

PCA alone is extremely powerful to reduce the number of variables; however, it is not a clustering algorithm and the group into clusters must be done with other techniques.

For example, Tanthanuch and colleagues applied FTIR microspectroscopy-supported by PCA and unsupervised hierarchical cluster analysis (UHCA) to identify specific spectral markers of the differentiation of murine embryonic stem cell (mESCs) and to distinguish them into different neural cell types [25]. In particular, focal plane array (FPA) - FTIR and SR-FTIR microspectroscopy measurements - performed on cell clumps and single cells respectively - allowed to obtain a biochemical fingerprint of different mESC developmental stages, namely embryoid bodies (EBs), neural progenitor cells (NPCs) and embryonic stem-derived neural cells (ESNCs). Interestingly, it should be noted that the results obtained on cell clumps and on single cells were found to be comparable, corroborating the FPA-FTIR results on cell clumps. The analysis of second derivative spectra enabled to highlight important spectral changes occurring during ES cell differentiation, mainly in the lipid CH_2 and CH_3 stretching region and in the protein amide I band. Noteworthy, these results overall indicated that during neural differentiation the cell lipid content increased significantly, likely reflecting modifications in cell membranes, whose lipid content is known to have a key role in neural cell differentiation and signal transduction. Moreover, changes in the profile of amide I band, mainly involving the alpha-helix component around 1650-1652 cm^{-1}, indicated an increased expression of alpha-helix reach protein in ESNCs compared with their progenitor cells, a result that could reflect the expression of cytoskeleton protein, crucial for the establishment of neural structure and function. These results were then strongly supported by PCA, that made it possible to disclose regions of the IR spectrum which most contributed to the spectral variance, namely amide I band and C-H stretching region.

Furthermore, the application of UHCA allowed to successfully discriminate and classify each stage of ESNCs differentiation, again considering the spectra in the spectral range mainly due to acyl chain vibrations and the extended region between 1750 and 900 cm^{-1}.

As discussed previously, PCA is frequently used for preliminary dimensionality reduction before further analyses, as LDA [21]. Indeed, a limit of using PCA alone is that it does not allow to obtain an unambiguous grouping of the data into clusters, requiring therefore the application of another analysis step able to reduce the intra-category variation while maximizing that inter-category [69]. The coupling, for instance, of PCA with LDA is a well established procedure which enables not only to classify the observations into groups but to quantify the importance of the single variables for this group separation. In this view, the advantage of LDA is that it makes it possible to reveal clusters, identifying objectively also the most contributory wavenumbers responsible for spectra discrimination [21, 58]. In particular, the application of PCA-LDA to spectroscopic investigation of complex biological systems proved to be a useful tool for the identification of spectral biomarkers of the process under investigation [7, 35, 69, 70, 71].

One outstanding work, worth to mention here, was done by Kelly and colleagues [70], where the authors showed how infrared spectroscopy and multivariate techniques can be used as a novel diagnostic approach for endometrial cancer screening. They first demonstrated how SR-FTIR microspectroscopy with subsequent PCA-LDA allows the clear segregation of different subtypes of endometrial carcinoma. However, the requirement of a particle accelerator impairs the use of endometrial spectroscopy as practical diagnostic application.

Recently, Taylor and colleagues applied ATR-FTIR spectroscopy supported by PCA-LDA analysis to interrogate endometrial tissues, employing in particular a conventional IR radiation source [72], showing that this approach, that can be applied directly to liquid or solid samples without further preparation, could provide a useful and simple objective test for endometrial cancer diagnosis.

Furthermore, in the work of Walsh and colleagues [69], ATR microspectroscopy has been successfully applied to the characterization of samples of exfoliative cervical cytology of different categories, with increasing severity of atypia. The spectral analysis was supported by PCA, with or without subsequent LDA, to verify if it was possible to discriminate among normal, low grade and high grade of exfoliative cytology. Indeed, important differences were found in the spectral range between 1500 and 1000 cm^{-1}, mainly due to proteins, glycoproteins, phosphates and carbohydrates. Noteworthy, the authors stressed that only the employment of the combined PCA-LDA allowed to maximize the inter-category variance, whilst reducing that intra-category. In particular, they found that the glycogen content strongly influenced the intra-category variance, while that inter-category resulted to be mainly due to protein and DNA conformational changes. In this view, FTIR microspectroscopy coupled with PCA-LDA could allow for an objective classification approach to class cervical cytology.

We should note that a delicate point of PCA-LDA is the choice of the principal components to be used as LDA input and, as described in the previous section about PCA, several ways have been developed to perform this task. Alternatively, the PLS method can be used instead of PCA [6, 73, 74]. For instance, Sandt and colleagues, using synchrotron infrared microspectroscopy coupled with PLS-DA, were able to characterize the metabolic fingerprint of induced pluripotent stem cells (iPSCs). In particular, they found that iPSCs are characterized by a chemical composition that leads to a spectral signature indistinguishable from that of embryonic stem cells (ESCs), but entirely different from that of the original somatic cells [6].

5.1. Ftir Microspectroscopy Supported By Pca-Lda For The Characterization Of Sn And Nsn Murine Oocytes

Recently, we applied FTIR microspectroscopy supported by PCA-LDA to the study of murine oocytes characterized by two different types of chromatin organization, namely surrounded nucleolus (SN) oocytes in which the chromatin is highly condensed and forms a ring around the nucleolus, and the not surrounded nucleolus (NSN) type where chromatin is dispersed and less condensed around the nucleolus [7, 75]. Interestingly, only SN oocytes are capable to complete the embryonic development after fertilization, while the NSN type, if fertilized, arrests at the two cell stage. To try to get new insights on the mechanisms that drive the different chromatin organization in the two kinds of oocytes, crucial for their embryonic development after fertilization, we studied the infrared absorption of single intact cells at different maturation stages, namely antral germinal vesicle (GV), metaphase I (MI, matured for 10 hours in vitro), and metaphase II (MII, matured for 20 hours in vitro).

Indeed, as we will show in the following, the FTIR spectra of the oocytes taken at the different maturation stages are very complex, since they provide information on different processes that were taking place simultaneously within the cells. For this reason, beside a fundamental visual inspection of the data, enabling the identification and assignment of the different spectral bands, it was crucial the application of PCA-LDA that made it possible to draw out the most significant spectral information responsible for the different cell behavior. Moreover, PCA-LDA allowed to identify the stage at which the separation between the SN and NSN oocytes took place, leading to their well distinct cell destinies.

As we discussed in paragraph 2, since the FTIR spectrum of cells is due to the overlapping contributes of the main biomolecules (see Figure 2), we analysed the second derivative spectra to identify the band peak positions and to assign them to the different biomolecule vibrational modes. The spectral analysis, strongly supported by PCA-LDA, allowed us to disclose the most important spectral differences between the two types of oocytes, at each maturation stage, that were found to occur mainly in the lipid and nucleic acid absorption regions, as we will discuss below. For a full discussion of the results see [7].

5.1.1. LIPID ANALYSIS

5.1.1.1. NSN oocytes

The analysis between 3050 and 2800 cm^{-1}, mainly due to the lipid carbon-hydrogen stretching vibrations [29], disclosed significant variations in the lipid content of NSN oocytes during their maturation up to MII. Indeed, besides an increase of the CH_2 band intensity up to MII, respectively at 2922 cm^{-1} and 2852 cm^{-1}, important changes concerned mainly the unsaturated fatty acid composition, as indicated by variations of the band between 3020 and 3000 cm^{-1} due to the olefinic group absorption. Indeed, as shown in Figure 3A, a single peak around 3013 cm^{-1} was present at GV and MI stages, while a splitting in two components at ~ 3016 cm^{-1} and at ~ 3010 cm^{-1} characterized the MII stage (see the inset of Figure 3A). These results could reflect important changes in membrane fluidity, which in turn could confer to the oocyte a different division ability after fertilization [8].

5.1.1.2. SN oocytes

SN oocytes were found to be characterized - during maturation up to MII - by a significant increase of the 2937 cm^{-1} component that could be likely due to cholesterol and/or phospholipids (Figure 3B) [76,77]. As discussed for NSN oocytes, the observed changes could reflect variations in the membrane properties, again highlighting the crucial role of lipids as markers of oocyte developmental competence [8, 78].

5.1.1.3. PCA-LDA analysis

The results obtained by the direct inspection of second derivative spectra were confirmed by PCA-LDA analysis performed on raw spectra. Firstly, the analysis was made on each type of oocyte taken at the different maturation stages. For the SN oocytes, the component carrying the highest discrimination weight resulted that at 2938 cm^{-1}, likely due to cholesterol and / or phospholipids [76, 77], in agreement with what found by the direct inspection of the spectra.

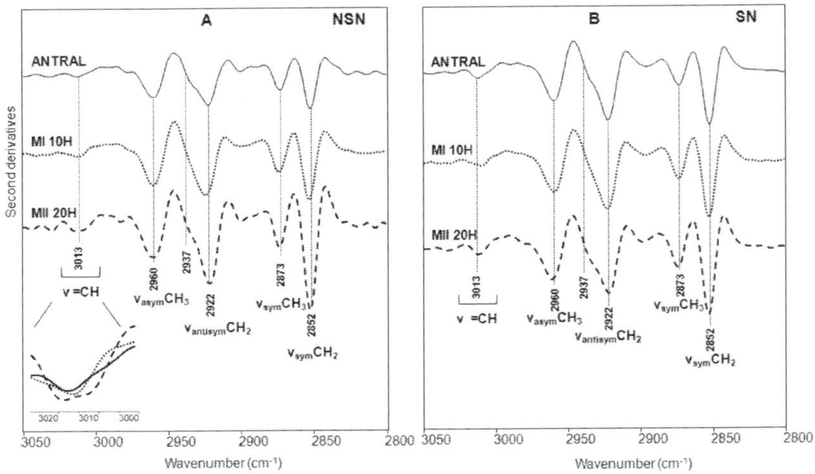

FIGURE 3. Second derivative absorption spectra of NSN (A) and SN (B) oocytes in the lipid absorption region. The second derivatives of the FTIR absorption spectra of single oocytes, measured at the antral (continuous line), MI 10 H (dotted line), and MII 20 H (dashed line) stages, are reported in the acyl chain absorption region, after normalization at the tyrosine peak (~1516 cm^{-1}). In the inset a magnification of the olefinic group band is shown.

Concerning the NSN oocytes, on the other hand, the wavenumbers with the highest discrimination weight were the 2922 cm^{-1}, due to the CH_2 stretching vibration, which increases up to MII, and the 3018 cm^{-1}, assigned to the olefinic group =CH of polyunsaturated fatty acids, whose absorption was observed to vary during the oocyte maturation.

We, then, compared the two types of oocyte at each maturation stage - as illustrated in Figure 4 - and we found that at the antral and MII stages the spectral components with the highest discrimination weight were those due to cholesterol and /or phospholipids, while at MI was that due to the olefinic group. Furthermore, to support the crucial role played by lipids in determining at some extent the oocyte developmental capacity, we should add that when we compared by PCA-LDA the spectra of the two oocyte types at the same maturation stage in the 1800-1500 cm^{-1} spectral range, dominated by the amide I and amide II absorption, the wavenumber with the highest discrimination weight was the 1739 cm^{-1}, due to the carbonyl stretching vibration of esters [7, 29].

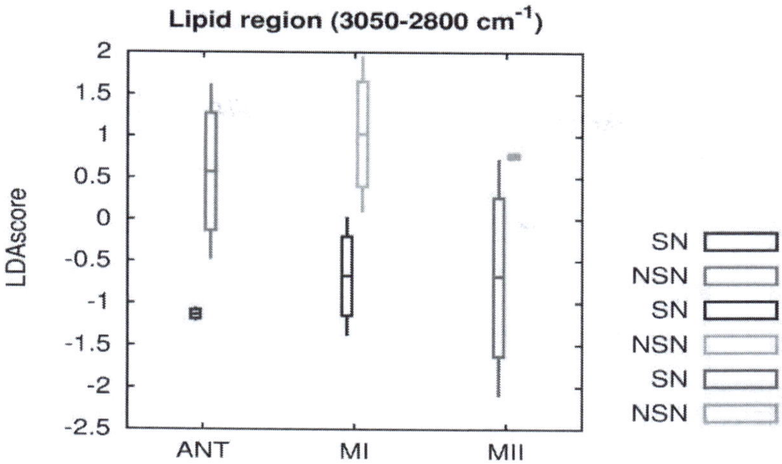

FIGURE 4. PCA-LDA analysis of SN and NSN oocytes in the lipid acyl chain absorption region (3050 – 2800 cm^{-1}). The separation between the two types of oocytes at each maturation stage is reported as average of PCA-LDA scores. The height of the boxes and the whiskers corresponds to 1 and 1.5 standard deviations from the mean values, respectively. The analysis has been performed on the measured spectra.

5.1.2. NUCLEIC ACID ANALYSIS

5.1.2.1. NSN oocytes

We then analyzed the nucleic acid IR response of NSN and SN oocytes during their maturation, exploring the spectral region between 1000 and 800 cm^{-1}, where RNA and DNA vibrational modes mainly occur [31, 32].

We found that NSN oocytes maintain, in all the studied stages, an appreciable transcriptional activity as indicated mainly by the simultaneous presence of the RNA ribose component around 921 cm^{-1} and of the DNA deoxyribose between 895-898 cm^{-1} - indicative of a DNA/RNA hybrid - whose relative intensities were seen to vary during maturation (see Figure 5A). In particular, the intensity of these two components is higher at the antral stage, while it decreases at MI, to increase again up to MII. These results were also supported by the response of the complex band between 980-950 cm^{-1}, mainly due to the CC stretching vibration of DNA backbone. Indeed, the profile of this band varies depending on the DNA structure that, in turn, could reflect a different nucleic acid activity. In particular, for the NSN oocytes we found that at the antral stage DNA is mainly in A-form - with a triplet at 975 cm^{-1}, 966 cm^{-1} and 951 cm^{-1} - typical of the DNA/RNA hybrid during transcription. At MI, the reduction of the 975 cm^{-1} and 966 cm^{-1} bands and the appearance of that at 969 cm^{-1} indicate that DNA is mainly in the B-form, suggesting a sort of transcriptional "stand by state", further supported by the reduction extent of the DNA/RNA hybrid, as discussed above. From this "stand by state" NSN oocytes seem to resume their transcriptional activity at MII, where a coexistence of DNA A and B forms was observed, as indicated by the increase of the ~ 975 cm^{-1} band and again in agreement with the simultaneous increase of the ribose (921 cm^{-1}) and deoxyribose (898 cm^{-1}) components.

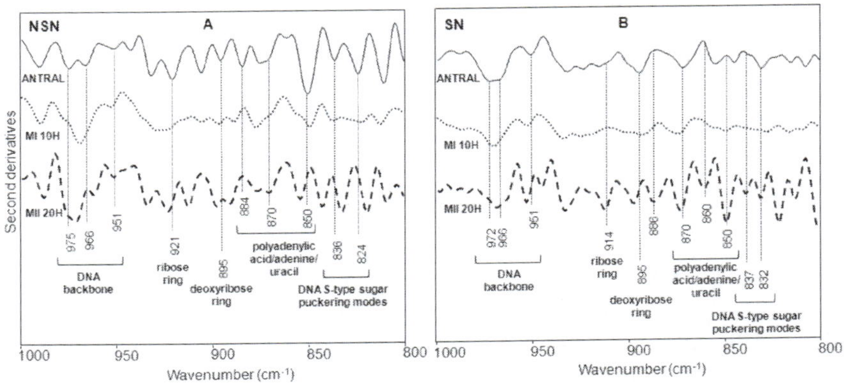

FIGURE 5. Second derivative absorption spectra of NSN (A) and SN (B) oocytes in the nucleic acid absorption region. The second derivatives of the FTIR absorption spectra of single oocytes, measured at the antral (continuous line), MI 10 H (dotted line), and MII 20 H (dashed line) stages, are reported in the 1000-800 cm^{-1} absorption region, after normalization at the tyrosine peak (~1516 cm^{-1}).

Furthermore, the analysis of the low frequency range, between 840-820 cm^{-1}, allowed us to obtain information on DNA methylation. In particular, in this spectral range, bands due to DNA S-type sugar puckering modes occur, which are sensitive to changes in the DNA sugar conformation induced by cytosine methylation [32]. The possibility to monitor changes in the profile of this spectral region in whole intact cells makes it possible, therefore, to obtain information on the variation of global DNA methylation in the CpG islands. In this way, we found that in the NSN oocytes DNA methylation was high at the antral stage, while it became very low, almost negligible at MII, in agreement with what found for the transcriptional activity pattern at the different maturation stages.

Finally, significant spectral differences were found between 890 and 850 cm^{-1}, where four different bands due to adenine and uracil vibrational modes occur (see Figure 5) [79]. Interestingly, the relative variation of these bands enables to monitor the mRNA polyadenylation extent, a crucial mechanism that regulates transcription. We found, in particular, that NSN oocytes were characterized during maturation by a low level of mRNA polyadenylation, being the polyadenylic acid band at 884 cm^{-1}absent at MII, while a new band at 854 cm^{-1} - likely due to adenine possibly not involved in polyA tail [80] – appeared. These results seem to suggest that an inadequate level of mRNA polyadenylation could preclude the possibility to resume meiosis, leaving the NSN oocytes in an unsuccessful transcriptional state.

5.1.2.2. SN oocytes

The analysis of SN oocytes (Figure 5B) in the spectral range between 1000 and 800 cm^{-1} led to very different results compared to NSN oocytes (see Figure 5A). Briefly, during all the studied maturation stages, the SN oocyte transcriptional activity was found to be maintained at lower levels than NSN oocytes, as revealed by the analysis of the CC stretching of the DNA backbone (980-950 cm^{-1}) and the monitoring of the ribose (~ 922 cm^{-1}) and deoxyribose (895-898 cm^{-1}) vibrations. These results were supported by the temporal evolution of the DNA methylation bands that suggested a partial CpG methylation at the antral and MI stages, which dramatically increased at MII, contrary to what observed for NSN oocytes.

Noteworthy, while no evidence of mRNA polyadenylation was observed for SN oocytes at the antral stage - as indicated by the absence of the two polyadenylic acid bands around 884 cm^{-1} and 860 cm^{-1} - starting from MI the adenine and uracile bands at 870 cm^{-1} and 850 cm^{-1} appeared, to then dramatically increase up to MII. These findings likely indicate that SN MII oocytes are characterized by an adequate level of maternal polyadenylated mRNAs, making them ready to sustain a proper embryo development, contrary to NSN oocytes.

5.1.2.3. PCA-LDA analysis

The above results overall indicate that the IR spectra of oocytes at different maturation stages are very informative in the nucleic acid absorption region, allowing to obtain information on several cell processes simultaneously, including transcriptional activity, DNA methylation, and RNA polyadenylation. For this reason, PCA-LDA analysis was crucial to disclose the most significant spectral response, enabling to identify the marker bands able to discriminate between the two kinds of oocytes.

Firstly, we analyzed the different maturation stages of each kind of oocyte. In particular, NSN oocytes displayed a segregation into three separated clusters, each corresponding to a maturation stage, with a classification accuracy of about 80%. Noteworthy, the wavenumber with the highest weight (1.0) was that around 880 cm^{-1}, due to polyadenylic acid, that, as revealed by second derivative analysis, was present only at the antral stage and disappeared upon maturation up to MII.

On the other hand, PCA-LDA analysis of SN oocytes led to an excellent discrimination accuracy (97%), with the wavenumbers with the highest discrimination weight at 817 cm^{-1} (1.0) and 859 cm^{-1}(0.83). While this last component is due to polyadenylic acid, the assignment of the 817 cm^{-1} band is not unequivocal, being due to overlapping contributions of DNA and polyadenylic acid.

The above results were then confirmed by the PCA-LDA analysis performed between 1400-1000 cm^{-1}, where contributions due to nucleic acids, such as sugar-phosphate vibrations, also occur [31]. In particular, for the NSN oocytes the wavenumber with the highest discrimination weight (1.0) was the 1305 cm^{-1}, which is due to free adenine, possibly not involved in polyadenylation [79]. In agreement with the temporal pattern of the adenine band at 870 cm^{-1}, discussed previously, the 1305 cm^{-1}component displayed a higher intensity at MII, confirming that an inadequate mRNA polyadenylation could preclude NSN oocytes from a successful embryonic development (see Figure 6).

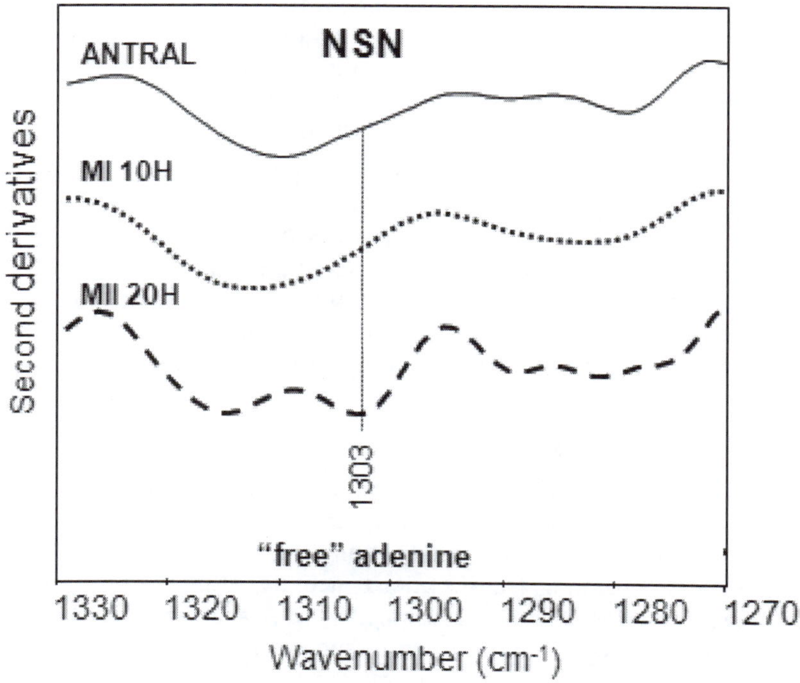

FIGURE 6. Second derivative absorption spectra of NSN oocytes in the absorption region of "free" adenine. The second derivatives of the FTIR absorption spectra of single NSN oocytes, measured at the antral (continuous line), MI 10 H (dotted line), and MII 20 H (dashed line) stages, are reported in the 1330-1270 cm^{-1} spectral range, where "free" adenine absorbs, after normalization at the tyrosine peak (~1516 cm^{-1}).

We then compared by PCA-LDA the two types of oocytes taken at the same maturation stage. As reported in Figure 7, we found the largest spectral distance at MI (92% classification accuracy), with the components carrying the highest discrimination weight due to A-DNA, likely reflecting differences in the transcriptional activity. In this view, MI stage could be considered a sort of crucial checkpoint, when some molecular rearrangements occur, deciding the oocyte fate.

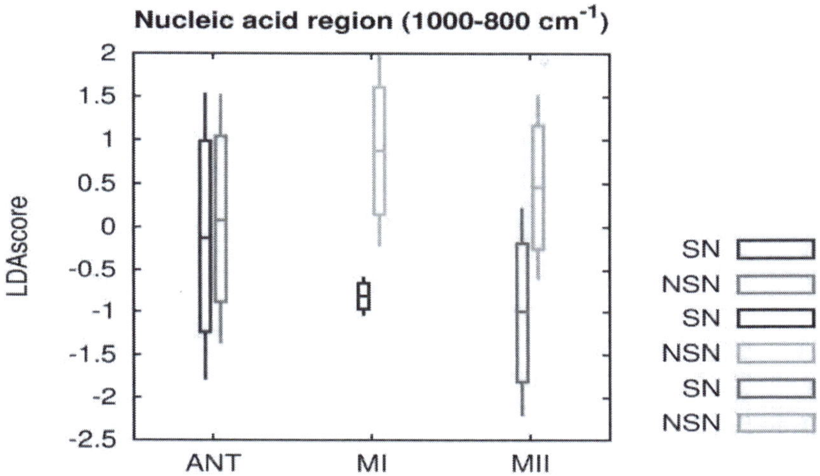

FIGURE 7. PCA-LDA analysis of SN and NSN oocytes in the nucleic acid absorption region (1000 - 800 cm^{-1}). The separation between the two types of oocytes at each maturation stage is reported as average of PCA-LDA scores. The height of the boxes and the whiskers corresponds to 1 and 1.5 standard deviations from the mean values, respectively. The analysis has been performed on the measured spectra.

These findings have been strongly supported by the comparison of the SN and NSN oocytes at each maturation stage, altogether. A very good discrimination accuracy (89%) was again found analyzing the nucleic acid absorption region, between 1000 and 800 cm^{-1}, that led to a clear cut separation into two groups (see Figure 8): one containing only the MII SN oocytes, and the other containing all the other SN and NSN stages. In particular, the wavenumbers carrying the highest discrimination weight were found at 926 cm^{-1} (1.00), due to ribose vibration, and at 855 cm^{-1} (0.97), assigned to adenine vibration, indicating again that differences in the temporal evolution and extent of transcription and polyadenylation play a crucial role in determining the different oocyte fate: the MII SN oocytes, with their proper content of maternal mRNAs polyadenylated, ready to support successfully the embryonic development; on the other hand, the MII NSN oocytes, with their mRNA lacking the appropriate polyadenylation, are kept in an unsuccessful transcriptional state.

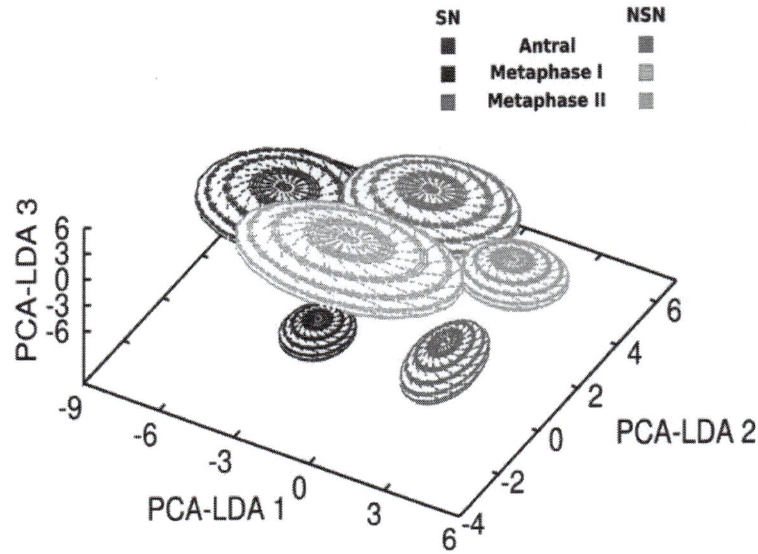

FIGURE 8. PCA–LDA analysis of SN and NSN oocytes in the nucleic acid absorption region. The PCA–LDA analysis has been carried out on measured FTIR absorption spectra obtained from SN and NSN oocytes at each maturation stage, between 1000 and 800 cm^{-1}. The semi-axes of ellipsoids in the 3D score plot correspond to two standard deviations of the data along each direction.

6. CONCLUSIONS

FTIR microspectroscopy has recently emerged as a powerful tool in biomedical research, thanks to the possibility of providing, in a non-invasive and rapid way, a chemical fingerprint of biological samples. In particular, being successfully applied to the study of complex biological systems, it makes it possible not only to characterize in situ biological processes, but also to provide a rapid diagnosis of several diseases, such as cancer and amyloid-based disorders.

We should, however, note that the intrinsic complexity of the IR response of biological systems - due to the overlapping absorption of the main biomolecules - requires the support of an appropriate multivariate analysis approach able to draw out the significant and non-redundant information contained in these highly dimensional data. Indeed, only a suitable combination of biospectroscopy and of multivariate analysis would provide robust and reliable results through the identification of specific biomarkers, an essential prerequisite for unbiased result interpretation [19, 20].

ACKNOWLEDGEMENT

D. A. is indebted to the University of Milano-Bicocca (I) for the supporting postdoctoral fellowship. P. M. acknowledges a postdoctoral fellowship from Italian Institute of Technology. S.M. D. acknowledges the financial support of the FAR (Fondo di Ateneo per la Ricerca) of the University of Milano-Bicocca (I).The authors wish to thank Carlo Alberto Redi and his collaborators at the University of Pavia (I) for the collaboration on murine oocyte maturation, and Antonino Natalello of the University of Milano-Bicocca (I) for helpful discussions.

REFERENCES

1. C Aksoy, F Severcan, Role of Vibrational Spectroscopy in Stem Cell ResearchSpectroscopy: An International Journal 2012273167184
2. D Ami, P Mereghetti, A Natalello, S. M Doglia, Fourier transform infrared microspectroscopy as a tool for embryonic stem cell studies. In: Atwood CS. (ed.) Stem Cells in Clinic and Research. Rijeka: InTech, 2011193218
3. P Heraud, E. S Nga, S Caine, Q. C Yu, C Hirst, R Mayberry, A Bruce, B. R Wood, D Mcnaughton, E. G Stanley, A. G Elefanty, Fourier transform infrared microspectroscopy identifies early lineage commitment in differentiating human embryonic stem cellsStem Cell Research 201042140147
4. J. W Chan, D. K Lieu, Label-free biochemical characterization of stem cells using vibrational spectroscopyJournal of Biophotonics2009211656668
5. M J Walsh, A Hammiche, T. G Fellous, J. M Nicholson, M Cotte, J Susini, N. J Fullwood, P. L Martin-hirsch, M. R Alison, F. L Martin, Tracking the cell hierarchy in the human intestine using biochemical signatures derived by mid-infrared microspectroscopyStem Cell Research2009311527
6. C Sandt, O Féraud, N Oudrhiri, M. L Bonnet, M. C Meunier, Y Valogne, A Bertrand, M Raphaël, F Griscelli, A. G Turhan, P Dumas, A Bennaceur-griscelli, Identification of spectral modifications

occurring during reprogramming of somatic cellsPLoS ONE 2012e30743.

7. D Ami, P Mereghetti, A Natalello, S. M Doglia, M Zanoni, C. A Redi, M Monti, FTIR spectral signatures of mouse antral oocytes: molecular markers of oocyte maturation and developmental competence.Biochimica et Biophysica Acta20111813612201229

8. B. R Wood, T Chernenko, C Matthäus, M Diem, C Chong, U Bernhard, C Jene, A. A Brandli, D Mcnaughton, M. J Tobin, A Trounson, O Lacham-kaplan, Shedding New Light on the Molecular Architecture of Oocytes Using a Combination of Synchrotron Fourier Transform-Infrared and Raman Spectroscopic MappingAnalytical Chemistry200880239065 9072

9. L Diomede, G Cassata, F Fiordaliso, M Salio, D Ami, A Natalello, S. M Doglia, A De Luigi, M Salmona, Tetracycline and its analogues protect Caenorhabditis elegans from β amyloid-induced toxicity by targeting oligomers. Neurobiology of Disease, 2010402424431

10. A Kuzyk, M Kastyak, V Agrawal, M Gallant, G Sivakumar, M Rak, Del Bigio MR, Westaway D, Julian R, Gough KM. Association among amyloid plaque, lipid, and creatine in hippocampus of TgCRND8 mouse model for Alzheimer diseaseThe Journal of Biological Chemistry20102854131 20231207

11. L. P Choo, D. L Wetzel, W. C Halliday, M Jackson, LeVine SM, Mantsch HH. In situ characterization of beta-amyloid in Alzheimer's diseased tissue by synchrotron Fourier transform infrared microspectroscopy.Biophysical Journal, 19967141 6721679

12. J Kneipp, L. M Miller, M Joncic, M Kittel, P Lasch, M Beekes, D Naumann, In situ identification of protein structural changes in prion-infected tissueBiochimica et Biophysica Acta20036393152158

13. G Bellisola, C Sorio, Infrared spectroscopy and microscopy in cancer research and diagnosisAmerican Journal of Cancer Research201221121

14. M. J Walsh, M. J German, M Singh, H. M Pollock, A Hammiche, M Kyrgiou, H. F Stringfellow, E Paraskevaidis, P. L Martin-hirsch, F. L Martin, I. R Microspectroscopy, potential applications in cervical cancer screening. Cancer Letters, 2007

15. C Petibois, G Déléris, Chemical mapping of tumor progression by FT-IR imaging: towards molecular histopathologyTrends in Biotechnology20062410455462

16. M. Z Kastyak-ibrahim, M. J Nasse, M Rak, C Hirschmugl, Del Bigio MR, Albensi BC, Gough KM. Biochemical label-free tissue imaging

with subcellular-resolution synchrotron FTIR with focal plane array detectorNeuroimage2012601376383

17. L. M Miller, P Dumas, From structure to cellular mechanism with infrared microspectroscopyCurrent Opinion in Structural Biology2010205649656

18. J Trevisan, P. P Angelov, P. L Carmichael, A. D Scott, F. L Martin, Extracting biological information with computational analysis of Fourier-transform infrared (FTIR) biospectroscopy datasets: current practices to future perspectives.The Analyst20121371432023215

19. J. G Kelly, J Trevisan, A. D Scott, P. L Carmichael, H. M Pollock, P. L Martin-hirsch, F. L Martin, Biospectroscopy to metabolically profile biomolecular structure: a multistage approach linking computational analysis with biomarkersJournal of Proteome Research201110414371448

20. L Wang, B Mizaikoff, Application of multivariate data-analysis techniques to biomedical diagnostics based on mid-infrared spectroscopyAnalytical and Bioanalytical Chemistry2008391516411654

21. F. L Martin, M. J German, E Wit, T Fearn, N Ragavan, H. M Pollock, Identifying variables responsible for clustering in discriminant analysis of data from infrared microspectroscopy of a biological sampleJournal of Computational Biology, 200714911761184

22. P Heraud, M. J Tobin, The emergence of biospectroscopy in stem cell researchStem Cell Research2009311214

23. S. G Kazarian, K. L Chan, Applications of ATR-FTIR spectroscopic imaging to biomedical samples.Biochimica et Biophysica Acta200617587858867

24. H-Y. N Holman, M. C Martin, W. R Mckinney, Tracking chemical changes in a live cell: Biomedical applications of SR-FTIR spectromicroscopy2003

25. W Tanthanuch, K Thumanu, C Lorthongpanich, R Parnpai, P Heraud, Neural differentiation of mouse embryonic stem cells studied by FTIR spectroscopyJournal of Molecular Structure2010189 EOF195 EOF

26. S Caine, P Heraud, M. J Tobin, D Mcnaughton, C. C Bernard, The application of Fourier transform infrared microspectroscopy for the study of diseased central nervous system tissueNeuroimage201259436243640

27. T Nakamura, J. G Kelly, J Trevisan, L. J Cooper, A. J Bentley, P. L Carmichael, A. D Scott, M Cotte, J Susini, P. L Martin-hirsch, S

Kinoshita, N. J Fullwood, F. L Martin, Microspectroscopy of spectral biomarkers associated with human corneal stem cellsMolecular Vision, 201016359

28. D Ami, A Natalello, A Zullini, S. M Doglia, Fourier transform infrared microspectroscopy as a new tool for nematode studies.FEBS Letters20045763297300

29. H. L Casal, H. H Mantsch, Polymorphic phase behaviour of phospholipid membranes studied by infrared spectroscopy.Biochimica et Biophysica Acta19847794381401

30. A Barth, Infrared spectroscopy of proteins.Biochimica et Biophysica Acta200717679l0731101

31. M Banyay, M Sarkar, A Gräslund, A library of IR bands of nucleic acids in solution.Biophysical Chemistry20031042477488

32. M Banyay, A Gräslund, Structural effects of cytosine methylation on DNA sugar pucker studied by FTIRJournal of Molecular Biology20023244667676

33. M Kačuráková, M Mathlouthi, FTIR and laser-Raman spectra of oligosaccharides in water: characterization of the glycosidic bond.Carbohydrate Research1996284145

34. P. T Wong, R. K Wong, T. A Caputo, T. A Godwin, B Rigas, Infrared spectroscopy of exfoliated human cervical cells: evidence of extensive structural changes during carcinogenesis.Proceedings of the National Academy of Sciences USA, 199188241098810992

35. D Ami, T Neri, A Natalello, P Mereghetti, S. M Doglia, M Zanoni, M Zuccotti, S Garagna, C. A Redi, Embryonic stem cell differentiation studied by FT-IR spectroscopy.Biochimica et Biophysica Acta20081783198106

36. S. O Konorov, H. G Schulze, N. J Caron, J. M Piret, M. W Blades, Turner RFB. Raman microspectroscopic evidence that dry-fixing preserves the temporal pattern of non-specific differentiation in live human embryonic stem cellsJournal of Raman Spectroscopy2011424576579

37. R Zhao, L Quaroni, A. G Casson, Fourier transform infrared (FTIR) spectromicroscopic characterisation of stem-like cell populations in human esophageal normal and adenocarcinoma cell lines. The Analyst, 201013515361

38. P Bassan, A Kohler, H Martens, J Lee, H. J Byrne, P Dumas, E Gazi, M Brown, N Clarke, P Gardner, Resonant Mie scattering (RMieS) correction of infrared spectra from highly scattering biological samplesThe Analyst20101352268277

39. H Susi, D. M Byler, Resolution-enhanced Fourier transform infrared spectroscopy of enzymes.Methods in Enzymology1986130290

40. I. W Levin, R Bhargava, Fourier transform infrared vibrational spectroscopic imaging: integrating microscopy and molecular recognition.Annual Review of Physical Chemistry200556429

41. P Dumas, L Miller, The use of synchrotron infrared microspectroscopy in biological and biomedical investigationsVibrational Spectroscopy, 2003323

42. J. E Katon, Infrared Microspectroscopy. A Review of Fundamentals and ApplicationsMicron, 1996275303314

43. L Eriksson, E Johansson, N Kettaneh-wold, J Trygg, C Wikstrom, S Wold, Multivariate and Megavariate Data Analysis Basic Principles and Applications. San Jose: Umetrics Academy; 2006

44. Manly BFJMultivariate Statistical Methods. London: Chapman & Hall/CRC press; 2004

45. M. R Hestenes, E Stiefel, Methods of Conjugate Gradients for Solving Linear SystemsJournal of Research of the National Bureau of Standards 195249409436

46. D Marquardt, An Algorithm for Least-Squares Estimation of Nonlinear ParametersSIAM Journal on Applied Mathematics 1963112431441

47. K Levenberg, A Method for the Solution of Certain Non-Linear Problems in Least Squares. Quarterly of Applied Mathematics 19442164

48. D E Goldberg, Genetic Algorithms in Search Optimization and Machine Learning.Boston: Addison-Wesley Longman Publishing; 1969

49. A C Rencher, Methods of Multivariate AnalysisHoboken: Wiley; 2002

50. T Nas, T Isaksson, T Fearn, T Davies, Multivariate Calibration and Classification. Chichester: NIR Publications; 2004

51. R. A Fisher, The use of multiple measurements in taxonomic problemsAnnals of Eugenics 19367179

52. T Fearn, 2002Discriminant analysis, In: Handbook of Vibrational Spectroscopy, Chalmers, J.M. & Griffiths, P.R. (eds.), New York: Wiley, 20862093

53. K Fukunaga, Introduction to Statistical Pattern RecognitionSan Diego: Academic Press Professional Inc.; 1990

54. C. M Bishop, Neural Networks for Pattern RecognitionNew York: Oxford University Press; 1995

55. P Jonathan, W. V Mccarthy, Roberts MIA. Discriminant Analysis With Singular Covariance Matrices. A Method Incorporating Cross-Validation And Efficient Randomized Permutation Tests. Journal Of Chemometrics, 1996104189213

56. S Rezzi, I Giani, K Héberger, D. E Axelson, V. M Moretti, F Reniero, and G Claude, Classification of Gilthead Sea Bream (Sparus aurata) from 1H NMR Lipid Profiling Combined with Principal Component and Linear Discriminant Analysis. Journal of agricultural and food chemistry 2007559963

57. L. S Skrobot, Castro VRE, Pereira RCC, Pasa VMD and Fortes ICP. Use of Principal Component Analysis (PCA) and Linear Discriminant Analysis (LDA) in Gas Chromatographic (GC) Data in the Investigation of Gasoline Adulteration. Energy & Fuels 2007213394

58. M. J Walsh, M. N Singh, H. M Pollock, L. J Cooper, M. J German, H. F Stringfellow, N. J Fullwood, E Paraskevaidis, P. L Martin-hirsch, F. L Martin, ATR microspectroscopy with multivariate analysis segregates grades of exfoliative cervical cytology.Biochemical and Biophysical Research Communications, 20073521213219

59. Pereira RCCSkrobot VL, Castro EVR, Fortes ICP, Pasa VMD. Determination of Gasoline Adulteration by Principal Components Analysis-Linear Discriminant Analysis Applied to FTIR SpectraEnergy & Fuels 20062010971102

60. K Héberger, E Csomós, andS Livia, Principal Component and Linear Discriminant Analyses of Free Amino Acids and Biogenic Amines in Hungarian Wines. Journal of Agricultural and Food Chemistry 2003518055

61. U. G Indahl, K. H Liland, T Naes, Canonical partial least squares- a unified PLS approach to classification and regression problems. Journal of Chemometrics 200923495

62. L Rieppo, J Rieppo, J. S Jurvelin, Saarakkala S: Fourier transform infrared spectroscopic imaging and multivariate regression for prediction of proteoglycan content of articular cartilagePloS ONE 2012e32344.

63. B Hemmateenejad, M Akhond, and F Samari, A comparative study between PCR and PLS in simultaneous spectrophotometric determination of diphenylamine, aniline, and phenol: Effect of wavelength selection, Spectrochimica Acta Part A: Molecular and Biomolecular Spectroscopy 200767958

64. A Krogh, What are artificial neural networks? Nature Biotechnology, 2008252195197

65. S Haykin, Neural Networks: A Comprehensive Foundation. Englewood Cliffs: Prentice Hall Inc.; 1999

66. A. J Bentley, T Nakamura, A Hammiche, H. M Pollock, F. L Martin, S Kinoshita, N. J Fullwood, Characterization of human corneal stem cells by synchrotron infrared micro-spectroscopy.Molecular Vision200713237

67. I Yousef, J Bréard, SidAhmed-Adrar N, Maâmer-Azzabi A, Marchal C, Dumas P, Le Naour F. Infrared spectral signatures of CDCP1-induced effects in colon carcinoma cellsThe Analyst20111362451625168

68. M. J German, A Hammiche, N Ragavan, M. J Tobin, L. J Cooper, S. S Matanhelia, A. C Hindley, C. M Nicholson, N. J Fullwood, H. M Pollock, F. L Martin, Infrared spectroscopy with multivariate analysis potentially facilitates the segregation of different types of prostate cellBiophysical Journal, 2006901037833795

69. M. J Walsh, M. N Singh, H. F Stringfellow, H. M Pollock, A Hammiche, O Grude, N. J Fullwood, M. A Pitt, P. L Martin-hirsch, F. L Martin, FTIR Microspectroscopy Coupled with Two-Class Discrimination Segregates Markers Responsible for Inter- and Intra-Category Variance in Exfoliative Cervical CytologyBiomarker Insights, 20083179

70. J. G Kelly, M. N Singh, H. F Stringfellow, M. J Walsh, J. M Nicholson, F Bahrami, K. M Ashton, M. A Pitt, P. L Martin-hirsch, F. L Martin, Derivation of a subtype-specific biochemical signature of endometrial carcinoma using synchrotron-based Fourier-transform infrared microspectroscopyCancer Letters20092742208217

71. M. J Walsh, T. G Fellous, A Hammiche, W. R Lin, N. J Fullwood, O Grude, F Bahrami, J. M Nicholson, M Cotte, J Susini, H. M Pollock, M Brittan, P. L Martin-hirsch, M. R Alison, F. L Martin, Fourier transform infrared microspectroscopy identifies symmetric PO(2)(-) modifications as a marker of the putative stem cell region of human intestinal crypts. Stem Cells, 2008261108118

72. S. E Taylor, K. T Cheung, I. I Patel, J Trevisan, H. F Stringfellow, K. M Ashton, N. J Wood, P. J Keating, P. L Martin-hirsch, F. L Martin, Infrared spectroscopy with multivariate analysis to interrogate endometrial tissue: a novel and objective diagnostic approachBritish Journal of Cancer, 20111045790797

73. C Chonanant, N Jearanaikoon, C Leelayuwat, T Limpaiboon, M. J Tobin, P Jearanaikoon, P Heraud, Characterisation of chondrogenic

differentiation of human mesenchymal stem cells using synchrotron FTIR microspectroscopyThe Analyst20111361225422551

74. K Thumanu, W Tanthanuch, Y Danna, S Anawat, L Chanchao, P Rangsun, H Philip, Spectroscopic signature of mouse embryonic stem cell-derived hepatocytes using synchrotron Fourier transform infrared microspectroscopy.Journal of Biomedical Optics2011057005 EOF

75. P Debey, M. S Szöllösi, D Szöllösi, D Vautier, A Girousse, D Besombes, Competent mouse oocytes isolated from antral follicles exhibit different chromatin organization and follow different maturation dynamics.Molecular Reproduction and Development19933615974

76. R Marty, N Soukpoé-kossi, C. N Charbonneau, D. M Kreplak, L Tajmir-riahi, HA. Structural characterization of cationic lipid-tRNA complexesNucleic Acid Research, 2009371551975207

77. J Liu, J. C Conboy, Structure of a gel phase lipid bilayer prepared by the Langmuir-Blodgett/Langmuir-Schaefer method characterized by sum-frequency vibrational spectroscopy.Langmuir, 2005212090919097

78. L Gentile, M Monti, V Sebastiano, V Merico, R Nicolai, M Calvani, S Garagna, C. A Redi, M Zuccotti, Single-cell quantitative RT-PCR analysis of Cpt1b and Cpt2 gene expression in mouse antral oocytes and in preimplantation embryosCytogenetic and Genome Research2004215 EOF221 EOF

79. G. P Zhizhina, E. F Oleinik, Infrared spectroscopy of nucleic acidsRussian Chemical Reviews1972413258280

80. G. N Ten, V. I Baranov, Manifestation of intramolecular proton transfer in imidazole in the electronic-vibrational spectrumJournal of Applied Spectroscopy2008752168173

Index